NATURAL KILLER T-CELLS: ROLES, INTERACTIONS AND INTERVENTIONS

NATURAL KILLER T-CELLS: ROLES, INTERACTIONS AND INTERVENTIONS

NATHAN V. FOURNIER
EDITOR

Nova Science Publishers, Inc.
New York

For permission to use material from this book please contact us:
Telephone 631-231-7269; Fax 631-231-8175
Web Site: http://www.novapublishers.com

NOTICE TO THE READER

The Publisher has taken reasonable care in the preparation of this book, but makes no expressed or implied warranty of any kind and assumes no responsibility for any errors or omissions. No liability is assumed for incidental or consequential damages in connection with or arising out of information contained in this book. The Publisher shall not be liable for any special, consequential, or exemplary damages resulting, in whole or in part, from the readers' use of, or reliance upon, this material. Any parts of this book based on government reports are so indicated and copyright is claimed for those parts to the extent applicable to compilations of such works.

Independent verification should be sought for any data, advice or recommendations contained in this book. In addition, no responsibility is assumed by the publisher for any injury and/or damage to persons or property arising from any methods, products, instructions, ideas or otherwise contained in this publication.

This publication is designed to provide accurate and authoritative information with regard to the subject matter covered herein. It is sold with the clear understanding that the Publisher is not engaged in rendering legal or any other professional services. If legal or any other expert assistance is required, the services of a competent person should be sought. FROM A DECLARATION OF PARTICIPANTS JOINTLY ADOPTED BY A COMMITTEE OF THE AMERICAN BAR ASSOCIATION AND A COMMITTEE OF PUBLISHERS.

Library of Congress Cataloging-in-Publication Data

Natural killer T-cells : roles, interactions, and interventions / Nathan V. Fournier (editor).
     p. ; cm.
  Includes bibliographical references and index.
  ISBN 978-1-60456-287-3 (hardcover)
  1. Killer cells. I. Fournier, Nathan V.
  [DNLM: 1. Killer Cells, Natural--immunology. 2. T-Lymphocytes--immunology. 3. Cytokines--biosynthesis. QW 568 N2856 2008]
  QR185.8.K54N38 2008
  616.07'99--dc22   2007052748

Published by Nova Science Publishers, Inc. ✦ *New York*

Contents

Preface

Natural killer T (NKT) cells are a heterogeneous group of T cells that share properties of both T cells and natural killer (NK) cells. Many of these cells recognize the non-polymorphic CD1d molecule, an antigen-presenting molecule that binds self- and foreign lipids and glycolipids. Upon activation, NKT cells are able to produce large quantities of interferon-gamma, IL-4, and granulocyte-macrophage colony-stimulating factor, as well as multiple other cytokines and chemokines (such as IL-2 and TNF-alpha). NKT cells seem to be essential for several aspects of immunity because their dysfunction or deficiency has been shown to lead to the development of autoimmune diseases (such as diabetes or atherosclerosis) and cancers. NKT cells have recently been implicated in the disease progression of human asthma. The clinical potential of NKT cells lies in the rapid release of cytokines (such as IL-2, IFN-gamma, TNF-alpha, and IL-4) that promote or suppress different immune responses.

Chapter 1 - NKT cells, originally characterized in mice as cells that express both a T cell receptor and NK1.1, have more recently been defined as cells that have an invariant Vα14-Jα18 (mouse) or Vα24-Jα18 (human) rearrangement and reactivity to α-galactosylceramide (α-GalCer) presented by the CD1d. The capacity of NKT cells to activate rapid cytokine expression has been exploited to manipulate the outcomes of autoimmunity and cancer.

NKT cells play an important role in pathogenesis of various diseases, including allergic asthma, inflammatory bowel diseases (IBD), mycobacterium tuberculosis, hepatitis, systemic lupus erythematosus (SLE), type 1 diabetes, and transplant tolerance. The authors discuss the function of NKT cells in these diseases, respectively.

NKT cells are innate-like T lymphocytes that recognize glycolipid antigens in the context of the MHC class I-related glycoprotein CD1d. Recent studies have identified multiple ways in which NKT cells can become activated during microbial infection. Mechanisms of CD1d-restricted antigen presentation are being unraveled, and a surprising connection has been made to proteins that control lipid metabolism and atherosclerosis. New studies have also provided important insight into the mechanisms that control effector cell differentiation of NKT cells and have revealed specialized functions of distinct NKT cell subsets. The authors' studies show that asthmatic NKT cells migrate from thymus, spleen, liver and bone marrow into blood vessels, and then concentrate in airway bronchi mucosa. This recruitment is dependent on high expression of CCR9 and engagement of CCL25/CCR9. NKT cells promote asthma in two different pathways (indirect and direct pathway).

These recent understanding of NKT cells performance in the development of asthma might unveil new therapy targets and management strategies for asthma. The authors' observations offer a potential explanation for high apoptotic sensitivity of NKT cells from active SLE, and provide a new insight into the mechanism of reduction of NKT cell number in SLE and understanding the association between NK T cell deficiency and autoimmune diseases. Nowadays, there is continued enthusiasm for the development of NKT cell-based therapies of human diseases.

NKT cells have now been shown to control various immune responses, including autoimmune, allergic, antitumor, and antimicrobial immune responses. For the definition of NKT cells, it is apparently inadequate that the simple classification of NKT cells as T cells also express NK receptors. Finally, these issues will be discussed in detail in this chapter.

Chapter 2 - OBJECTIVE: Lifestyle factors including cigarette smoking and alcohol consumption are associated with mortality and morbidity due to cancers of various organs. It is well documented that natural killer (NK) cells provide host defense against tumors and viruses. In order to explore whether lifestyle and a forest bathing trip (natural aromatherapy) affects human NK and lymphokine-activated killer (LAK) activities, NK and $CD56^+/CD3^+$ NKT cells, intracellular levels of perforin, granzymes, and granulysin, and the underlying mechanism of any effects, the authors have conducted a series of investigations to clarify this issue.

METHODS: Fifty-four to one hundred and fourteen healthy male subjects, aged 20-60 years, from a large company in Osaka, Japan were selected with informed consent. The subjects were divided into groups with good, moderate, and poor lifestyles according to their responses on a questionnaire regarding eight health practices (cigarette smoking, alcohol consumption, sleeping hours, working hours, physical exercise, eating breakfast, balanced nutrition, and mental stress). Peripheral blood was taken, and NK and LAK activities were measured by ^{51}Cr-release assay, numbers of NK, T, and $CD56^+/CD3^+$ NKT cells, and perforin-, granulysin-, and granzymes A/B-expressing cells in peripheral blood lymphocytes (PBL) were measured by flow cytometry. In the forest bathing trips, 12 healthy male subjects aged 35-56 years were selected with informed consent. The subjects experienced three-day/two-night trips to a forest area and to a city without forests, respectively. Blood was sampled on the second and third days during the trips, and on days 7 and 30 after the forest bathing trip. Similar measurements were made before the trips on normal working days as a control. NK activity was measured by ^{51}Cr-release assay, numbers of NK, T, and $CD56^+/CD3^+$ NKT cells, and perforin-, granulysin-, and granzymes A/B-expressing cells in PBL were measured by flow cytometry.

RESULTS: Subjects with good or moderate lifestyles showed significantly higher NK and LAK activities, higher numbers of NK cells, and perforin-, granulysin-, and granzymes A/B-expressing cells and a significantly lower number of T cells in PBL than subjects with a poor lifestyle. Among the eight health practices, cigarette smoking, physical exercise, eating breakfast, working hours, and balanced nutrition significantly affected the numbers of NK, T, and $CD56^+/CD3^+$ NKT cells, and perforin-, granulysin-, and granzymes A/B-expressing cells, and alcohol consumption significantly affected the number of granzyme A-expressing cells and NKT cells. On the other hand, mental stress and sleeping hours had no effect on these parameters. The forest bathing trip significantly increased NK activity, the number of NK cells, and perforin-, granulysin- and granzymes A/B-expressing cells, and the positive rate of NKT cells. In contrast, a city tourist visit did not show any effects on the above parameters.

CONCLUSIONS: Taken together, these findings indicate that good lifestyle practices and forest bathing trips significantly increased human NK and LAK activity, and the numbers of NK and NKT cells, and perforin-, granulysin-, and granzymes A/B-expressing cells in PBL.

Chapter 3 - Organophosphorus pesticides (OPs) are widely used throughout the world as insecticides in agriculture and eradicating agents for termites around homes. The main toxicity of OPs is neurotoxicity, which is caused by the inhibition of acetylcholinesterase. OPs also affect immune response including effects on antibody production, IL-2 production, T cell proliferation, Th1/Th2 cytokine profiles, decrease of CD5 cells, and increases of CD26 cells and autoantibodies, inhibitions of lymphokine-activated killer and cytotoxic T lymphocytes activities. However, there have been few papers investigating the mechanism of OP-induced inhibition of natural killer (NK) activity. This study reviews the effect of organophosphorus compounds on NK activity and the mechanism of organophosphorus compound-induced inhibition of NK activity. It has been reported that NK cells induce cell death in tumor or virus-infected target cells by two main mechanisms. The first mechanism is direct release of cytolytic granules that contain the pore-forming protein perforin, several serine proteases termed granzymes, and granulysin by exocytosis to kill target cells, which is called the granule exocytosis pathway. The second mechanism is mediated by the Fas ligand (Fas-L)/Fas pathway, in which FasL (CD95L), a surface membrane ligand of the killer cell cross links with the target cell's surface death receptor Fas (CD95) to induce apoptosis of the target cells. The authors had previously found that organophosphorus compounds including OPs significantly inhibit human and animal NK activity both in vitro and in vivo. Moreover, they found that OPs inhibit NK activity by at least the following three mechanisms: 1) OPs impair the granule exocytosis pathway of NK cells by inhibiting the activity of granzymes, and by decreasing the intracellular level of perforin, granzymes and granulysin, which was mediated by inducing degranulation of NK cells and by inhibiting the transcript of mRNA of perforin, granzyme A and granulysin. 2) OPs impair the FasL/Fas pathway of NK cells, as investigated by using perforin-knockout mice, in which the granule exocytosis pathway of NK cells does not function and only the FasL/Fas pathway remains functional. 3) OPs induce apoptosis of NK cells, which is at least partially mediated by activation of intracellular caspase-3.

Chapter 4 – Natural killer T (NKT) cells are one of the most mysterious immune system cells represented by unusual regulatory T lymphocytes which co-express some NK cell markers, and have the capacity to recognize glycolipid antigens in the context of MHC I-like molecule – CD1d via their invariant T cell receptor (TCR). In response to TCR ligation, NKT cells rapidly produce large amounts of both pro-inflammatory T helper (Th) 1 type cytokines and anti-inflammatory Th2 type cytokines. This paradoxical property has made it difficult to predict the consequences of NKT cell activation in vivo but has nonetheless caused much speculation that NKT cells play a grand role in immunoregulation.

The physiological function of NKT cells in antitumor immunity may be multifaceted. Despite accumulated experimental evidence that has supported the critical role of NKT cells in promoting effective tumor immunosurveillance, several studies suggest a contrary role for NKT cells consisted in dramatic suppression of acquired antitumor immunity. Recent studies discovered a veiled enigma of controversial role of NKT cells for tumor immunosurveillance and revealed that at least two subsets of CD1d-restricted NKT cells exist: type I NKT cells that can substantially influence function of other various cell types, particularly DC, NK cells, conventional CD4$^+$ and CD8$^+$ T cells, all contributing to the antitumor immunity, and type II NKT cells that have a potency to repress antitumor immune responses.

Immunotherapy strategies aimed at augmentation of CD1d-restricted type I NKT cell numbers or increasing of their activity are currently a major focus. Many latest experimental data and clinical trials clearly suggested that targeting NKT cells may provide a novel effective strategy for immunotherapy of incurable patients with malignancies; moreover frequency and/or function of NKT cells may be directly related to cancer disease prognosis.

Chapter 5 - Natural killer T (NKT) cells are innate immune cells that recognize lipid moieties presented by MHC -like CD1 molecules. The ability to secrete large amounts of cytokines rapidly upon activation has attracted considerable attention to their functions. Within the last decade it has become clear that NKT cells can steer immune responses towards either inflammation or tolerance. This ability is mostly attributed to the variety of different cytokines distinct NKT cell subsets can secrete and employ to generate a wide variety of effects on immunity, including control of infectious diseases, allergic responses, and cancer, as well as tolerance in autoimmunity and transplantation settings. However, our understanding of the activation mechanisms of NKT cells suffers from the limited knowledge of the origin and identity of endogenous glycolipids that are responsible for the development and activation of NKT cells. The use of alpha-galactosylceramide (a-GalCer), an exogenous and semi-artificial glycolipid, elicits the secretion of various cytokines including those with antagonistic biological activities, leading to considerable confusion about which factors account for the effect of NKT cells in their different roles. This chapter summarizes current knowledge of the impact of NKT cells in different peripheral tolerance models and discusses possible mechanisms which control the biological outcomes, immunity or tolerance.

Chapter 6 - The anti-leukemic potential of natural killer (NK) cells and their competence in regulating normal and possibly neoplastic hematopoietic precursors have over the years raised considerable interest. The role of NK cells in the immunosurveillance against tumor growth is well documented. Previous studies have shown that leukemic blasts may be susceptible to the lytic action of lymphokine-activated killer cells. More recently, NK clones of donor origin have been established in the post-transplant period from HLA-mismatched hematopoietic stem cell transplanted patients. NK clones were capable of killing recipients' leukemic cells, in the absence of graft-versus-host disease (GVHD). The authors have recently demonstrated the possibility of expanding cytotoxic NK cells with killing activity against autologous and allogeneic blasts from acute lymphoid leukemia (ALL) and acute myeloid leukemia (AML) patients in complete remission (CR). The cytolytic properties of this expanded cell population have been further confirmed in vivo in a NOD/SCID mouse tumor model that showed a consistent reduction of AML load after adoptive transfer of autologous expanded interleukin (IL)-2- and IL-15-activated NK cells into tumor-transplanted mice. These results are of particular interest if we consider the high rate of relapse that characterizes the clinical course of leukemia patients. These effectors expanded ex vivo may be used for vaccination programs aimed at controlling or eradicating minimal residual disease in leukemic patients in clinical and hematologic CR. Alternatively, this population of cytotoxic cells may be utilized in the setting of allografted patients, both in the haploidentical or HLA-matched scenario. Clinical protocols based on autologous or allogeneic NK cells expanded under Good Manufacturing Practice conditions appear feasible, particularly considering that the infusion of NK cells should induce very limited toxicity and no or a very low risk of GVHD.

Chapter 7 - HIV infection and its outcome are complex because there is great heterogeneity not only in clinical presentation, but also because of incomplete clinical information of markers of immunodeficiency and in measurements of viral loads. Also, there many gene variants that control not only viral replication but immune responses to the virus; it has been difficult to study the role of the many AIDS restricting genes (ARGs) because their influence varies depending on the ethnicity of the populations studies and because the cost to follow infected individuals for many years. Nevertheless, at least genes of the major histocompatibility locus (MHC) such as HLA alleles have been informative to classify infected individuals following HIV infection, progression to AIDS and long-term-non-progressors (LTNP). For example, progressors could be defined as up to 5 years, up to 11 years or as the authors describe in this report up to 15 years from infection, and LTNP could be individuals with normal CD4+ T cell counts for more than 15 years with or without high viral loads. In this review, the authors emphasized that in the studies of ARGs the HLA alleles are important in LTNP; HLA-B alleles influence the advantage to pathogens to produce immune defense mediated by CD8+ T cells (cognate immunuity). The main point of their report is that contrary to recent reports claiming that this dominant effect was unlikely due to differences in NK activation through ligands such as HLA-Bw4 motif, they believe that cognate immunity as well as innate inmmunity conferred by NK cells are involved. The main problem is that HLA-Bw4 alleles can be classified according the aminoacid in position 80. Isoleucine determines LTNP, which is a ligand for 3DS1. Such alleles did not include HLA-B*44, B*13 and B*27 which have threonine at that position. Other studies have not considered the fact that in addition to the NK immunoglobulin receptors, NK receptors can be of the lectin like such as NKG2A/HLA-E to influence the HIV infection outcome. HLA-Bw4 as well as HLA-Bw6 alleles can be classified into those with threonine or methionine in the second position of their leader peptides. These leader peptides are ligands for NKG2A in which methionine influences the inhibitory role of NKG2A for killing infected targets. Functional studies have not been done as well as studies of these receptors in infected individuals. However, analyses of the leader peptides of HLA-B alleles in published reports suggested that threonine in the second position can explain the importance of HLA-B*57, B*13, B*44 as well as certain Bw6 alleles in LNTP. In addition, the authors analyzed the San Francisco database that was reported and found that the association of HLA-B alleles with LNTP or with progressors can be due to the presence of threonine or methionine in their second position. Therefore, studies of outcome of HIV infection should include not only mechanisms of cognate immunity mediated by peptides and CD8+ T cells, but also NK receptors of two types, NKG2A as well as 3DS1. The authors propose that the SCID mouse should be used to understand mechanisms mediated by many of the ARGs, especially the importance of thymus derived cells as well as NK receptor interactions with their ligands in this experimental animal transplanted with human stem cells, thymus or NK cells obtained from individuals of known HLA genotypes.

Chapter 8 - Natural killer T (NKT) cells are a subset of regulatory lymphocytes that have been implicated in the regulation of autoimmune processes. The major histocompatibility complex class I-like CD1d glycoprotein is a member of the CD1 family of antigen presenting molecules, and is responsible for the selection of NKT cells. CD1d presents a number of ligands to NKT or other CD1d-restricted T cells, including glycolipids from a marine sponge, bacterial glycolipids, normal endogenous glycolipids, tumor-derived phospholipids and glycolipids, and non-lipid molecules. Some of these glycolipid/phospholipid ligand-CD1d complexes have been crystallized, revealing their tertiary structures. Most available data is on alpha configuration ligands. Recently, β-glycolpids have emerged as a family of possible ligands for this subset of regulatory lymphocytes. The presentation of many of these molecules can have immune-potentiating effects, acting as adjuvant against infections or promoting more rapid clearance of certain viruses. β-glycolpids can also be protective against autoimmune diseases or cancer; they can also be deleterious. In this review the authors discuss the potential use of these ligands against immune-mediated disorders.

In: Natural Killer T-Cells: Roles, Interactions and Interventions   ISBN: 978-1-60456-287-3
Editor: Nathan V. Fournier, pp. 1-49   © 2008 Nova Science Publishers, Inc.

Chapter 1

NKT Cell-Associated Diseases and Related Treatments

Tan Jinquan, Li Li, Wang Li, Xie Luokun,
Peng Tao and Zhou Rui*

Department of Immunology, Wuhan University
School of Basic Medical Sciences, Wuhan University
Wuhan 430071, P. R. China

I. Abstract

NKT cells, originally characterized in mice as cells that express both a T cell receptor and NK1.1, have more recently been defined as cells that have an invariant Vα14-Jα18 (mouse) or Vα24-Jα18 (human) rearrangement and reactivity to α-galactosylceramide (α-GalCer) presented by the CD1d. The capacity of NKT cells to activate rapid cytokine expression has been exploited to manipulate the outcomes of autoimmunity and cancer.

NKT cells play an important role in pathogenesis of various diseases, including allergic asthma, inflammatory bowel diseases (IBD), mycobacterium tuberculosis, hepatitis, systemic lupus erythematosus (SLE), type 1 diabetes, and transplant tolerance. We will discuss the function of NKT cells in these diseases, respectively.

NKT cells are innate-like T lymphocytes that recognize glycolipid antigens in the context of the MHC class I-related glycoprotein CD1d. Recent studies have identified multiple ways in which NKT cells can become activated during microbial infection. Mechanisms of CD1d-restricted antigen presentation are being unraveled, and a surprising connection has been made to proteins that control lipid metabolism and atherosclerosis. New studies have also provided important insight into the mechanisms that control effector cell differentiation of NKT cells and have revealed specialized functions of distinct NKT cell subsets. Our studies show that asthmatic NKT cells migrate from thymus, spleen, liver and bone marrow into blood vessels, and then concentrate in airway bronchi mucosa. This recruitment is dependent on high expression

* E-mail address: jinquan_tan@hotmail.com or, E-mail address: jinquan_tan@whu.edu.cn Tel: 0086 27 68758600, Fax: 0086 27 68758600. (Corresponding author)

of CCR9 and engagement of CCL25/CCR9. NKT cells promote asthma in two different pathways (indirect and direct pathway).

These recent understanding of NKT cells performance in the development of asthma might unveil new therapy targets and management strategies for asthma. Our observations offer a potential explanation for high apoptotic sensitivity of NKT cells from active SLE, and provide a new insight into the mechanism of reduction of NKT cell number in SLE and understanding the association between NK T cell deficiency and autoimmune diseases. Nowadays, there is continued enthusiasm for the development of NKT cell-based therapies of human diseases.

NKT cells have now been shown to control various immune responses, including autoimmune, allergic, antitumor, and antimicrobial immune responses. For the definition of NKT cells, it is apparently inadequate that the simple classification of NKT cells as T cells also express NK receptors. Finally, these issues will be discussed in detail in present chapter.

II. Introduction of the Historical and Current Research on NKT Cells

The term "NKT cells" was first published in 1995 and was used broadly to define a subset of mouse T cells that share some characteristics with NK cells, which express NK cell-associated receptors, such as CD161 in human and NK1.1 in mice [1, 2]. Thus early definitions of NKT cells were based on co-expression of the NK-cell marker NK1.1 and αβTCR [3, 4]. With the advent of CD1d tetramers, it became clear that not all NKT cells express NK1.1. Although the exact definition of NKT cells have not been identified, people are used to naming the subset of cells as Natural killer T (NKT) cells that share properties of natural killer cells and conventional T cells.

1. Classification, Subsets and Distribution of NK T Cell (Table 1)

NKT cells were divided into two major subpopulations, namely CD1d-restricted NKT cells and CD1d-independent NKT cell subsets (NKT-like cells) [5]. The former recognizes lipid Antigen presented by CD1d and has a highly skewed TCR repertoire. In contrast to CD1d-restricted NKT cells, which have been intensively studied, very little is known about CD1d-independent NKT cells [6, 7]. CD1d is an MHC-like, but genetically unlinked family of β2-microglobulin–associated molecules that has recently been shown to present lipid antigens to αβT cells [8, 9]. At least two classes of CD1d- restricted NKT cells have been defined. Type I NKT cells, also known as invariant or semi-invariant NKT cells (iNKT cells) express an invariant T-cell receptor α-chain (TCRα; Vα14-Jα18 in mice, Vα24-Jα18 in human) in combination with certain Vβ chains (Vβ8.2, Vβ7 or Vβ2 in mice, Vβ11 in humans). Type I NKT cells recognize the glycosphingolipid antigen α-galactosylceramide (α-GalCer) and are best defined using CD1d tetramers loaded with this antigen (α-GalCer–CD1d tetramers). Type II NKT cells are also CD1d dependent, but they express more diverse TCR Vα chains and do not recognize α-Gal-Cer, Much less is known about type II NKT cells [2, 10].

In this article we will focuse on Type I NKT cells (invariant or semi-invariant NKT cells (iNKT cells): Vα14-Jα18 NKT cells (mice), Vα24-Jα18 NKT cells (human)) ,herein referred to simply as NKT cells.

NKT cells, including a CD4$^+$, a CD8$^+$ and a CD4$^-$CD8$^-$ (DN) population, also express many markers commonly associated with the NK cell lineantigene including NK1.1, CD122, CD16, DX5 and Ly49 family members. They also bear an activated phenotype, *i.e.* CD44high, CD69+, CD62Low, and unlike most conventional T cells, NKT cells express CD38 and Ly6C [11,12].

Three distinct subsets (CD4$^+$, CD8$^+$ and DN) of NKT cells are differentially distributed in a tissue-specific fashion. CD8$^+$ NKT cells are present in all tissues but thymus, and are highly enriched for CD8 α$^+$ β$^-$ cells. Developmentally, most CD4$^+$ and DN NKT cells are thymus dependent, in contrast to CD8$^+$ NKT cells, and are also present amongst recent thymic emigrants in spleen and liver [11].

Table 1. Classification of NKT and NKT-like cells.

	Type I NKT cells (classical)	**Type II NKT cells (non-classical)**	**NKT-like cells**
CD1d dependent	Yes	Yes	No
α-GalCer reactive	Yes	No	No
TCR α-chain	Vα14-Jα18 mice Vα24-Jα18 humans	Diverse	Diverse
TCR β-chain	Vβ8.2, Vβ2 and Vβ7 (mice) β11 (humans)	Diverse, but some Vβ8.2(mice)	Diverse
NK 1.1 (CD 161)	+(resting mature) -/low (immature or post-activation)	+/-	+
Subsets	CD4$^+$ and DN(mice) CD4$^+$ CD8$^+$ and DN (humans)	CD4$^+$ and DN(mice)	CD4$^+$ CD8$^+$ and DN
Distribution	CD4$^+$ and DN: thymus/spleen/liver CD8$^+$: all tissues but thymus	thymus spleen liver	CD4$^+$ and DN: thymus/spleen/liverC D8$^+$: all tissues but thymus

2. Development of NKT Cells

The development of NKT cells is poorly understood [13]. For many years, the developmental origin of NKT cells was a matter of debate. Some studies suggested that they developed very early in ontogeny, independently of the thymus, and before the appearance of conventional T cells. However, there is now overwhelming evidence that NKT cells are a thymus-dependent population [14-16]. There is also convincing evidence that NKT cells segregate from conventional T-cell development at the double positive (DP;CD4$^+$CD8$^+$) thymocyte stantigene in thymic cortex [17,18,19].

The NKT cells pool in thymus contains immature (NK1.1$^-$) and mature (NK1.1$^+$) subsets that represent distinct linear stantigenes of a linear developmental pathway. Stuart P. Berzins et al have determined that mature NK1.1$^+$NKT cells are retained by the thymus to form an extremely long-lived resident population capable of rapid and prolonged production of IFN-γ and IL-4 [20]. Dale I Godfrey etal reported NKT cells arised in the thymus from a common precursor pool of CD4$^+$CD8$^+$ double positive (DP) thymocytes that have undergone random T-cell receptor (TCR) gene rearrangement and expression. Expression of a TCR that binded with appropriate avidity to self-peptide–MHC class II or I molecules on thymic epithelial cells (TECs) leaded to the positive selection of conventional CD4$^+$ or CD8$^+$ T cells, respectively. Thymocytes that expressed a TCR that interacts with CD1d bound to self glycolipid, expressed by other DP thymocytes, enter the NKT-cell lineantigene. Once selected, NKT-cell precursors undergo a series of differentiation steps that ultimately results in the NKT-cell pool. Immature NKT cells in the thymus were NK1.1 negative, and at least four distinct immature stantigenes have been defined through differential expression of CD24, CD44 and DX5. NKT cells then undergo an actively regulated maturation step, which ultimately results in a range of functional and phenotypic changes, including expression of NK1.1. Most NKT cells that emigrated from thymus do so as immature cells in the periphery. Some mature thymic NKT cells also migrated to the periphery, but many remained as long-term thymus-resident cells. Both immature and mature NKT cells included CD4$^+$ and CD4$^-$ subsets. Although they were functionally distinct, the developmental relationship between these subsets was not fully understood [10].

There are some molecules such as SAP, DOCK2, AP-3 required during the NKT cells development. SAP may integrate a set of long-standing yet seemingly disparate observations characterizing NKT cell development [21]. DOCK2 is required for positive selection of Vα14 NKT cells in a cell-autonomous manner, thereby suggesting that avidity-based selection also governs development of this unique subset of lymphocytes in the thymus [22]. The Adaptor Protein AP-3 is required for CD1d-mediated antigen presentation of glycosphingolipids and development of NKT Cells [23].

3. NKT Cells and DCs

Some people showed pathogen recognition by Toll-like receptors (TLRs) on dendritic cells (DCs) leaded to DC maturation and the initiation of adaptive immunity. But Christian et al. comment on the evidence that these pathways are TLR independent and have the potential to respond to infection, malignancy, and immunotherapy [25]. Recent studies have shown that innate lymphocytes-natural killer (NK), natural killer T (NKT), and γδT cells-also trigger DC maturation. This interaction in turn expands and activates innate lymphocytes and initiates adaptive T cell immunity [24]. Satoshi et al. demonstrate the mechanisms of generation of regulatory dendritic cells (DCs) by stimulation of Vα14 NKT cells in vivo. After repeated injection of α-GalCer into mice, splenic DCs acquired properties of regulatory DCs in IL-10-dependent fashion, such as immature phenotypes and increased IL-10 but reduced IL-12 production. The unique cytokine profile in these DCs appears to be regulated by ERK1/2 and I-BNS. These DCs also show an ability to suppress the development of experimental allergic encephalomyelitis by generating IL-10-producing regulatory CD4$^+$ T cells in vivo [26].

Mouse NKT cells are innate cells activated by glycolipid Antigens and play important roles in the initiation and regulation of immune responses. Someone indicate that during the course of murine schistosomiasis, NKT cells exhibit an activated phenotype and that following schistosome egg encounter in the liver, hepatic NKT cells produce both IFN-γ and IL-4 in vivo. They also report that schistosome egg-sensitized dendritic cells (DCs) activate, in a CD1d-dependent manner, NKT cells to secrete IFN-γ and IL-4 in vitro. Interestingly, transfer of egg-sensitized DCs promotes a strong Th2 response in recipient wild-type mice, but not in mice that lack NKT cells. Engagement of TLRs in DCs is not necessary for NKT cell stimulation in response to egg-sensitized DCs, suggesting an alternative pathway of activation. Therefore, they propose that self, rather than parasite-derived, CD1d-restricted ligands are implicated in NKT cell stimulation. Taken together, their data show for the first time that helminths can activate NKT cells to produce immunoregulatory cytokines in vivo, enabling them to influence the adaptive immune response [27, 28].

4. Specialized Functions of Distinct NKT Cell Subsets

NKT cells have diverse immune regulatory functions including activating cells involved in Th1 and Th2 type immune activities. Most previous studies had investigated the functions of NKT cells as a whole but more recent evidence indicates the distinct functional properties of NKT cell subpopulation. Lin H et al. suggest that NKT subpopulations differ in their abilities to stimulate other immune cells. They reported that the CD4$^+$CD8$^-$ NKT cells demonstrated substantially greater stimulatory activities on CD4$^+$ T cells, NK cells, and B cells than other NKT cell subsets. The CD4$^-$CD8$^+$ NKT cells showed the greatest activity on CD8$^+$ T cells, and were the only NKT cell subset that activated these immune cells. The CD4$^-$CD8$^-$ NKT cells showed moderate stimulatory activity on CD4$^+$ T cells and the least activity on other immune cells [29, 30].

III. NKT Cells and Diseases

NKT cells are a thymus-dependent T-cell subset, but are developmentally and functionally distinct from mainstream CD4$^+$ and CD8$^+$ T cells [18]. NKT-cell recognition of glycolipid antigens presented by CD1d leads to a cascade of events involving cytokines and co-stimulatory molecules that result in the activation of antigen-presenting cells (APCs) and many other bystander cells, including NK, T and B cells. The diversity and extent of cytokine production can have a broad range of effects, ranging from enhanced cell-mediated immunity Th1-type response to suppressed cell-mediated immunity Th2-type response [31, 32]. In most experimental cases, the response is advantantigeneous to the host. However in some cases, the response can be deleterious [Fig. 1] [10,33,34]. Upon activation these NKT cells rapidly produce high levels of interleukin IL-2, IFN-γ, TNF-α, and IL-4 [35]. NKT cells are involved in immediate immune responses, tumor rejection, immune surveillance and control of autoimmune diseases, such as diabetes, lupus, atherosclerosis, and allergen-induced asthma. NKT cells have also been shown to regulate viral infections in vivo, and control tumor growth [36].

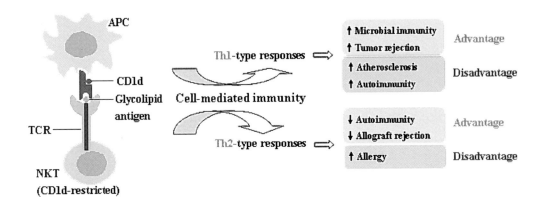

Figure 1. NKT-cell recognition of glycolipid antigens presented by CD1d leads to a cascade of events involving cytokines and co-stimulatory molecules that result in the activation of antigen-presenting cells (APCs) and many other bystander cells. The diversity and extent of cytokine production can have a broad range of effects, ranging from enhanced cell-mediated immunity (Th1-type responses) to suppressed cell-mediated immunity (Th2-type responses). In most experimental cases, the response is advantageous to the host, however, in some cases, the response can be deleterious.

1. Allergic Asthma

Allergic asthma is a multifaceted syndrome consisting of eosinophil-rich airway inflammation, bronchospasm, and airway hyper-responsiveness (AHR), in which conventional CD4[+] T cells producing IL-4/IL-13 appear to play an obligatory pathogenic role. In a mouse model of asthma, activation of pulmonary IL-4/IL-13 producing NKT cells is sufficient for the development of airway hyper-reactivity (AHR) [37, 38].

NKT cells are the prominent manipulator in asthma development. Asthmatic NKT cells migrate from thymus, spleen, liver and bone marrow into blood vessels, and then concentrate in airway bronchi mucosa. Recruitment of key effector cells may control disease severity, responsiveness to current therapies and the airway remodelling associated with the disease. Recent studies have discovered Chemokines and their receptors may play an important role for leukocyte trafficking in allergic inflammation [39], and their antantigenonists have the hope of preventing inflammatory cell recruitment to the airway and perhaps as a consequence affect the resolution of airway remodeling [40]. In Kim YK etals' study, they suggested that CCR5 and RANTES/CCL5 were key contributors to the development and maintenance of chronic fungal asthma in a mouse model [41].Their data regarded the role of CCR5 gene deletion (CCR5*D32 allele) in pathogenesis of asthma. It have been known RANTES play an important role in the pathogenesis of allergic asthma. However, Sandford AJ et al indicated that the CCR5*D32 allele was not a genetic risk factor for the development of asthma and did not influence disease severity. The CCR5*D32 allele did not influence RANTES production in heterozygous state [42]. While Fainaru O et al reported that the Runx3 transcription factor was a key regulator of lineantigene-specific gene expression in several developmental pathways and mutations in RUNX3 might be associated with increased sensitivity to asthma development. Which in a Runx3 knockout mice modle, CCR7 expression was enhanced, resulting in increased migration of alveolar DCs to the lung-draining lymph node, which was associated with development of asthma-like features, including increased serum IgE, hypersensitivity to inhaled bacterial lipopolysaccharide, and methacholine-induced airway hyperresponsiveness [69].

Respiratory administration of glycolipid antigens that specifically activate NKT cells (α-Galcer and a *Sphingomonas* bacterial glycolipid) rapidly induce AHR and inflammation typically associated with protein allergen administration. Everett H. Meyer et al suggested that NKT cells responding to glycolipid antigens as well as CD4$^+$ T cells responding to peptide antigens might be synergistic in the induction of AHR. Although in some cases, each might induce AHR independently. These demonstrated that NKT cells not only produced cytokines influencing adaptive immunity but also function as critical effector cells in AHR [43].

Allergic bronchopulmonary aspergillosis (ABPA) is a rare variant of severe asthma resulting from hypersensitivity to Aspergillus fumigatus (Asp f) present in the airways. Garcia G et al analyzed the expression of six chemokine receptors (CCR3, CCR4, CCR8, CCR5, CXCR3 and CXCR4) on total blood CD4$^+$ T cells and Asp f-specific T cells in ABPA patients, and reported the down-regulation of CCR4 and CXCR3 in Asp f-specific T cells after allergen exposure seem to be a feature of ABPA patients which required further evaluation [44].

In our study, we demonstrate NKT cells (esp.Vα24$^+$ i NKT cells) selectively express CCR9 at high frequency in asthma. (The correlation and mechanism will be described in detail in part IV).

2. Inflammatory Bowel Diseases (IBD)

NKT cells have been proposed to be both protective and pathogenic to inflammatory bowel diseases (IBD). On one hand, recent studies have shown that these cells are involved in the maintenance of mucosal homeostasis. On the other, NKT cells play a pathogenic role in human ulcerative colitis. Similar contrasting data have been seen in murine IBD models [45].

Crohn disease (CD) has been clearly identified as a Th1 inflammation, the immunopathogenesis of its counterpart, IBD and ulcerative colitis (UC), remains enigmatic. But Fuss IJ et al provided firm evidence that UC was associated with atypical Th2 response mediated by non-classical IL-13-producing NKT cells which have cytotoxic potential for epithelial cells [46].

3. Mycobacterium Tuberculosis

Most researchers concluded Mycobacterium tuberculosis (MTB) was a leading cause of mortality worldwide from an infectious antigen. However, the possible contribution of NKT cells to resistance to Mycobacterium tuberculosis infection remains unclear.

Many findings indicated that NKT cells might be an important component of anti-tuberculosis immunity, they thought NKT cells recognize mycobacterial antigens and contributed to anti-MTB immunity in mouse models [47]. In Dieli F etals' study, they characterized the NKT cell population following infection with Mycobacterium bovis bacillus

Calmette-Guérin (BCG), and supported a regulatory role for NKT cells in the course of BCG infection as a regulator of the balance between protection and pathogenesis [48]. Siobhán C etal showed that a rare population of CD4⁻CD8⁻CD3⁺αβ⁺γδ⁻NK1.1⁻T cells expanded into effector cells, and then differentiated into long-lived memory cells after infection, these cells potently and specifically inhibited the growth of the intracellular Mycobacterium or Francisella tularensis Live Vaccine Strain (LVS) in macrophages in vitro, promoted survival of mice infected with these organisms, and generate adoptively transferred immunity to F. tularensis LVS [49].

In tuberculosis, almost every cell types, such as CD4⁺ and CD8⁺ T cells, NK cells, NKT cells and pro-inflammatory macrophages can act in a synergic way to control the growth and multiplication of Mycobacterium [51]. Someone indicated Interleukin-10 suppressesed NK cell but no NKT cell activation during bacterial infection [50].

4. Hepatitis

NKT cells are abundant in the liver of mice and human. Recent studies have revealed that NKT cells participate in some types of liver injuries, such as concanavalin A-induced T cell-mediated hepatitis and malaria hepatitis. Here we focus on some factor-involving liver damantigenes and NKT cell-mediated liver injuries [52].

Someone suggested that engagement of P2X(7)Rs on NKT cells inhibited naive, while stimulating activated cells, resulting in suppression or stimulation of autoimmune hepatitis. Micromolar concentrations of nicotinamide adenine dinucleotide (NAD) induced a rapid increase of annexin V staining in NKT cells in vitro, a response that required expression of P2X(7)Rs. Consistent with this result, treatment of mice with NAD caused a temporary decrease of NKT cells in the liver and protected from Con A- and α-GalCer-induced hepatitis, both of which required functional NKT cells. Resistance to liver injury was associated with decreased cytokine production by NKT cells in NAD-treated mice. In contrast, when NAD was injected into Con A- or α-GalCer-primed mice, liver injury was exacerbated and cytokine production by NKT cells was increased. This effect was caused by P2X(7)R-mediated stimulation of activated NKT cells. In antigenreement, mice lacking P2X(7)Rs on lymphocytes suffer reduced liver injury, and animals lacking ADP-ribosyltransferase, the enzyme that used NAD to attach ADP-ribosyl groups to cell surfaces, were also resistant to Con A-induced hepatitis [62] .

Upon Con A administration, hepatic NKT cells rapidly up-regulate cell surface FasL expression and FasL-mediated cytotoxicity. At the same time, NKT cells undergo apoptosis leading to their rapid disappearance in the liver. These results implicated FasL expression on liver NKT cells in the pathogenesis of Con A-induced hepatitis, suggesting a similar pathogenic role in human liver diseases such as autoimmune hepatitis [63].

Someone indicated that activated NKT cells negatively regulate liver regeneration of HBV-tg mice in the PHx model. In a model of human HBV infection, partial hepatectomy (PHx) markedly triggered NKT cell accumulation in the hepatectomized livers of HBV-tg mice, simultaneously with enhanced IFN-γ production and CD69 expression on hepatic NKT cells at the early stantigene of liver regeneration. The impairment of liver regeneration in HBV-tg mice was largely ameliorated by NKT cell depletion, but not by NK cell depletion. Blockantigene of NKT cell interaction considerably alleviated NKT cell activation and their inhibitory effect on regenerating hepatocytes. Neutralization of IFN-γ enhanced bromodeoxyuridine incorporation in HBV-tg mice after PHx, and IFN-γ mainly induced hepatocyte cell cycle arrest. Adoptive transfer of NKT cells from regenerating HBV-tg liver, but not from normal mice, could inhibit liver regeneration in recipient mice. Dong, Z etal demonstrated in the HBV transgenic mouse model that activation of NKT by a single injection of α-GalCer induced IFN-γ production, which was sufficient in controlling viral replication without further lymphocyte infiltration. However, NKT cells may also cause liver damantigene, as demonstrated in murine hepatitis models induced by concanavalin A or LPS plus IL-12. NKT cells disappeared from livers following their activation, probably as a result of apoptosis. The newly emerging NKT cells were biased to produce only IL-4 upon restimulation. Because of their ability to produce large quantities of IL-4, it was initially thought that NKT cells might be essential for inducing Th2 immune responses. However, the experimental induction of these types of responses in $CD1d^{-/-}$ and $β_2M^{-/-}$ mice (which lack NKT cells) showed that these cells were not absolutely necessary for these responses [53].

Recently, Sun R et al. [60] have reported IL-6/STAT3 inhibited NKT cells via targeting $CD4^+$ T cells and consequently prevents T cell-mediated hepatitis. While De Lalla C et al. [61] reported that NKT cells responded to progressive liver damantigene caused by chronic hepatitis virus infection, and suggested that these cells, possibly triggered by the recognition of CD1d associated with viral- or stress-induced lipid ligands, contribute to pathogenesis of cirrhosis by expressing a set of cytokines involved in the progression of fibrosis. In Biburger M etals' study, it showed Liver injury induced by α-GalCer injection into C57BL/6 mice was accompanied by intrahepatic caspase-3 activity but appeared independent thereof α-GalCer injection also induces pronounced cytokine responses, including TNF-α, IFN-γ, IL-2, IL-4, and IL-6. Cytokine neutralization revealed that, unlike Con A-induced hepatitis, IFN-γ is not only dispensable for α-GalCer-induced hepatotoxicity but even appears to exert protective effects. In contrast, TNF-α was clearly identified as an important mediator for hepatic injury in this model. Whereas intrahepatic Kupffer cells are known as a pivotal source for TNF-α in Con A-induced hepatitis, they were nonessential for α-GalCer-mediated hepatotoxicity [55].

NKT cells patrol liver sinusoids to provide intravascular immune surveillance CXCR6 contributes to liver-based immune responses by regulating their expression. CXCR6-deficient mice exhibited a selective and severe reduction of CD1d-reactive NKT cells in liver and decreased susceptibility to T-cell-dependent hepatitis [58]. There was a similar model showed that increase in hepatic NKT cells in Leukocyte cell-derived chemotaxin 2-deficient mice contributes to severe concanavalin A-induced hepatitis. Leukocyte cell-derived chemotaxin 2 (LECT2) was originally identified for its possible chemotactic activity against human neutrophils in vitro. It was a 16-kDa protein that was preferentially expressed in the liver. Its homologues have been widely identified in many vertebrates. Current evidence suggests that LECT2 may be a multifunctional protein like cytokines. However, the function of LECT2 in vivo remains unclear. Interestingly, the hepatic injury was exacerbated in LECT2 ($^{-/-}$) mice upon treatment with Con A, possibly because of the significantly higher expression of IL-4 and FasL. These results suggested that LECT2 might regulate the homeostasis of NKT cells in the liver and might be involved in the hepatitis pathogenesis [56].

Someone indicated persistent infection with hepatitis C virus (HCV) was presumably explained by a deficient immune response to the infection. NS3, a nonstructural, HCV-encoded protein, induced a prolonged release of oxygen radicals from mononuclear and polymorphnuclear phantigenocytes. In turn, the NS3-activated phantigenocytes induced dysfunction and/or apoptosis in three major subsets of lymphocytes of relevance to defense against HCV infection: $CD3^+/56^-$ T cells, $CD3^-/56^+$ NK cells, and $CD3^+/56^+$ NKT cells [59].

Some studies to note it is important that NKT cell activation is accompanied by NK cell activation, which is essential for the effector functions of NKT cells [54]. Someone suggest CD44 also plays an important role in the regulation and elimination of immune cells in liver. CD44 deficiency on hepatic mononuclear cells leads to reduced activation-induced apoptosis which results in increased liver damantigene. In SEB-induced liver injury study, injection of staphylococcal enterotoxin B (SEB, a bacterial superantigen) into d-galactosamine-sensitized CD44 wild-type mice lead to a significant increase in CD44 expression on liver T cells, NK cells, and NKT cells [57].

5. Autoimmune Disease

A numerical and functional deficit of NKT cells has been reported to be associated with pathogenesis of autoimmune diseases, and the immunoregulatory function of NKT cells is crucial for prevention of autoimmunity. They are potent producers of immunoregulatory cytokines, and have been implicated in several different autoimmune diseases in mice and humans, including Type I diabetes, experimental autoimmune encephalomyelitis, systemic lupus erythematosus and scleroderma [64].

The prototypical NKT cell Antigen α-GalCer is not present in mammalian cells, and little is known about the mechanism of NKT cell recruitment and activation. Someone suggested that up-regulation of CD1d expression during inflammation was critical for maintaining T cell homeostasis and preventing autoimmunity [65]. NKT cells have been reported to be deficient in patients with a variety of autoimmune syndromes and in certain strains of autoimmune mice. In addition, injection of mice with α-GalCer activated these T cells and ameliorates autoimmunity in different disease models. Thus, deficiency and reduced function in NKT cells were considered to be risk factors for such diseases [66]. Someone suggested that the reduced numbers of NKT cells in patients with autoimmune diseases might be due to an inadequate amount of α-GalCer-like natural ligands (i.e., adequate in only 48.6% of patients) for the induction of NKT cells in vivo, or due to dysfunction of NKT cells (in 51.4% of patients) [67]. Both the frequency and the absolute number of $CD161^+CD8^+$ T cells were decreased in peripheral blood of patients suffering SLE, MCTD, SSc and PM/DM, which suggested that there was also an abnormality of NKT cells in $CD8^+$ population [68].

Autoantibody production and lymphadenopathy are common features of systemic autoimmune disease. Targeted or spontaneous mutations in the mouse germline have generated many autoimmune models with these features. Importantly, the models have provided evidence for gene function in prevention of autoimmunity, suggesting indispensable roles for many genes in normal immune response and homeostasis.

1) Systemic Lupus Erythematosus (SLE)

Patients with SLE, scleroderma, diabetes, multiple sclerosis, and rheumatoid arthritis have lower numbers of peripheral NKT cells [70, 71].

Forestier C reported that the development of systemic lupus erythematosus in (New Zealand Black (NZB) x New Zealand White (NZW))F(1) mice was paradoxically associated with an expansion and activation of NKT cells. Although young (NZB x NZW)F(1) mice had normal levels of NKT cells, NKT cells expanded with antigene and became phenotypically and functionally hyperactive. Activation of NKT cells in (NZB x NZW)F(1) mice in vivo or in vitro with α-GalCer indicated that the immunoregulatory role of NKT cells varied over time, revealing a marked increase in their potential to contribute to production of IFN-γ with antigene and disease progression. This evolution of NKT cell function during the progression of autoimmunity may have important implications for the mechanism of SLE and for therapies using NKT cell antigenonists [66]. Recent studies have shown pleiotropic roles of SAP in T, B, and NK cell activations and NKT cell development [72]. MRL/lpr, a mutant strain mice are known to stably exhibit systemic lupus erythematosus-like diseases. However, the mutant mice barely displayed autoimmune phenotypes, though the original defect in Fas expression was unchanged. Linkantigene analysis using (mutant MRL/lpr x C3H/lpr) F2 mice demonstrated a nucleotide insertion that caused loss of expression of small adaptor protein, signaling lymphocyte activation molecule (SLAM)-associated protein (SAP). SAP is known to be a downstream molecule of SLAM family receptors and to mediate the activation signal for tyrosine kinase Fyn.

2) Type I Diabetes

Type I diabetes is an autoimmune disease due to the destruction of insulin-producing pancreatic β cells, and NKT cells play a key regulatory role in type I diabetes. The absence of NKT cells correlates with exacerbation of type I diabetes, whereas an increased frequency and/or activation of NKT cells prevents β-cell autoimmunity [73].Various mechanisms are involved in the protective effect of NKT cells, especially in the NOD mouse model [74-78]. Analysis of NKT cell regulatory function in the NOD mouse has revealed that NKT cells inhibit the development of type I diabetes by impairing the differentiation of anti-islet T cells into Th1 effector cells, and Diabetes is still prevented by NKT cells in the absence of functional IL-4, IL-10, IL-13, and TGF-β [79-82]. In addition, glycolipid antigenonists of NKT cells have been successful in preventing diabetes in mice, raising enthusiasm for the development of NKT cell-based therapies for TID [83]. Cain JA et al [176] showed that NKT cells efficiently dampened the action of diabetogenic CD4[+] T cells, and did so in an indirect manner by modifying the host environment. Moreover, the NKT cell-containing population modifies the host via production of IFN-γ that was necessary for driving the inhibition of diabetogenic T cells in vivo.

In several models of type I diabetes, increasing the number of NKT cells prevented the development of disease. However, so far, published data are contradictory in regard to the role of NKT cells in Type I diabetes mellitus (TIDM). In Kis J et al' study the TIDM NKT cell cytokine profile markedly shifted to Th1 direction. There was no difference in the frequency of NKT cells in PBMC among the different patient groups. The decrease in the CD4[+] population of NKT cells and their Th1 shift indicates dysfunction of these potentially important regulatory cells in TIDM [85]. Griseri T etal showed NKT cells exacerbate type 1 diabetes induced by CD8 T cells. They indicated CD8[+] T cells play a crucial role in the pathogenesis of diabetes and high frequency of NKT cells promoted severe insulitis and exacerbated diabetes. type I diabetes was induced by the transferring CD8[+] T cells specific for the influenza virus hemantigenglutinin into recipient mice of which β pancreatic cells specifically expressing hemantigenglutinin antigen. Analysis of diabetogenic CD8[+] T cells showed that NKT cells enhance their activation, expansion, and differentiation into effector cells producing IFN-γ. This first analysis of the influence of NKT cells on diabetogenic CD8[+]

T cells revealed that NKT cells not only fail to regulate but in fact exacerbate the development of diabetes. Thus, NKT cells can induce opposing effects dependent on the model of type I diabetes being studied. This prodiabetogenic capacity of NKT cells should be taken into consideration when developing therapeutic approaches based on NKT cell manipulation [84].

The NOD mouse has been proved to be a relevant model of insulin-dependent diabetes mellitus, closely resembling human disease. However, it is unknown whether this strain presents a general bias toward Th1-mediated autoimmunity or remains capable of mounting complete Th2-mediated responses. Someone proved that autoimmune diabetes-prone NOD mice could also give rise to enhanced Th2-mediated responses and might thus provide a useful model for the study of common genetic and cellular components, including NKT cells that contribute to both asthma and type I diabetes[86]. Araujo LM et al showed that NOD mice had the capacity of developing a typical Th2-mediated disease, namely experimental allergic asthma. This was the first evidence that autoimmune diabetes-prone NOD mice can also give rise to enhanced Th2-mediated responses and might thus provide a useful model for the study of common genetic and cellular components, including NKT cells that contribute to both asthma and type I diabetes. They even developed a stronger Th2-mediated pulmonary inflammatory response than BALB/c mice, a strain that showed a typical Th2 bias in this model. Thus, after allergen sensitization and intra-nasal challenge, the typical features of experimental asthma were exacerbated in NOD mice, including enhanced bronchopulmonary responsiveness, mucus production and eosinophilic inflammation in the lungs as well as specific IgE titers in serum. These hallmarks of allergic asthma were associated with increased IL-4, IL-5, IL-13 and eotaxined production in the lungs, as compared with BALB/c mice. Notwithstanding their quantitative and functional defect in NOD mice, NKT cells contributed to antigengravate the disease, since in OVA-immunized CD1d$^{-/-}$ NOD mice, which were deficient in this particular T cell subset, airway eosinophilia was clearly diminished relative to NOD littermates [87].

6. Transplant Tolerance

Allogeneic bone marrow transplantation is a curative treatment for leukemia and lymphoma, NKT cells are thought to regulate immune responses and to play important roles in the induction of allograft tolerance. Host NKT cells can prevent graft-versus-host disease (GVHD) and permit graft antitumor activity after bone marrow transplantation. Someone indicated the role for CD8$^+$ T cells versus generalized MHC class I-restricted antigen presentation in islet allograft rejection and tolerance promoted allograft prolongation and it was a major role for host MHC class I antigen presentation to promote islet allograft survival. Adoptive transfer of small numbers of DX5$^+$ cells alleviates GVHD in a murine model of semiallogeneic bone marrow transplantation, indicating a potential role for NKT cells [88].

It is well-documented that certain chemokines or their receptors play important roles in the graft rejection. However, the roles of chemokines and their receptors in the maintenance of transplantation tolerance remain unclear. Xiaofeng Jiang et al demonstrated the unique role of CXCL16 and CXCR6 molecules in the maintenance of cardiac allograft tolerance mediated by NKT cells. In a mouse transplant tolerance model, the expression of CXCL16 was up-regulated in the tolerated allografts, and anti-CXCL16 mAb inhibited accumulation of NKT cells. In vitro experiments further showed that blocking CXCL16/CXCR6 interaction significantly affected not only chemotaxis but also NKT cell adhesion [89].

There are still lacking data supporting a role for NKT cells in the maintenance of human tissue-specific tolerance. Galante NZ et al are interested in studying NKT cell frequency in kidney transplant recipients and its correlation with graft function. They observed that total number of NKT cells did not differ significantly among transplant patients when compared to normal controls, although specific-subsets seem to be more frequent in determined events [90]. While Frequency of NKT cells were decreased after living-donor liver transplantation (LDLT). Although the contribution of those subsets to the tolerant state remains elusive, the results may provide important clues for reliable indicators of tolerance after LDLT [91]. Harantigenuchi K et al also reported that the number of NKT cells in bone marrow transplantation (BMT) recipients with acute GVHD was lower than that in patients without acute GVHD, and both the CD4[+] and CD4[-] Valpha24[+] NKT subsets were significantly reduced after hematopoietic stem cell transplantation. With regard to chronic GVHD, BMT recipients with extensive GVHD had significantly fewer NKT cells than other patients. Furthermore, the number of CD4[+] Valpha24[+] NKT cells was also significantly reduced in patients with chronic extensive GVHD [92, 93].

Neonatal tolerance is a very interesting phenomenon, because even allogeneic skin grafts are not rejected in these mice at the adult stantigene. However, the underlying mechanism remains unclear. Kawamura H et al concluded with additional data from a cell-transfer experiment and a splenectomy experiment, CD8[+] NKT cells in the liver of tolerant mice might be intimately associated with the neonatal tolerance phenomenon [94].

7. Tumor Immunity

Distinct NKT cell subsets have activities of tumor rejection. The DN subset of NKT cells from the liver were better good at reject tumor cells than were the CD4[+] liver derived subset and both the CD4[+] and DN subsets from the thymus or the spleen. Type II NKT cells also participate in the suppression of tumor rejection, type II NKT cell is sufficient for downregulation of tumor immunosurveillance. The importance of immunoregulatory T cells has become increasingly apparent.

Both CD4+CD25+T cells and CD1d-restricted NKT cells have been reported to down-regulate tumor immunity in mouse tumor models. However, the relative roles of both T cell populations have rarely been distinguished in the same tumor models. In addition, CD1d-restricted NKT cells have been reported to play a critical role not only in downregulation of tumor immunity but also in the promotion of the immunity. However, the explanation for these apparently opposite roles in different tumor models remains unclear. Masaki etal showed that in four mouse tumor models in which CD1d-restricted NKT cells suppression of tumor immunity, depletion of CD4[+]CD25[+] T cells did not induce enhancement of immune surveillance. Surprisingly, among the two subpopulations of CD1d-restricted NKTcells, type I NKT cells were not necessary for the immune suppression. These unexpected dates may resolve the paradox regarding the role of CD1d-restricted NKT cells in regulation of tumor immunity. Type II NKT cells may be sufficient for negative regulation, whereas protection has been found to be mediated by α-galactosylceramide-responsive type I NKT cells [95].

IV. Insights into the Roles of NKT Cells in Various Diseases

NKT cells have been identified as an important component of the immune system that has the capacity both of augmenting beneficial host immunity and of preventing harmful autoimmunity [96]. Whether the apparent differences in NKT response patterns depends on variations in NKT antigens and/or on presence of specific subsets of mucosal NKT cells [97]. Most recent studies have elucidated clearly CD1d-restricted antigen-presentation mechanisms and development of various diseases based on this pathway upon NKT activation.

Therefore, we will introduce CD1d processing and antigen presentation firstly, and then describe our group discoveries in detail; finally, elucidate the mechanism of NKT cells in several diseases.

1. CD1d-Restricted Antigen-Presentation Mechanisms

CD1d molecules are synthesized in the endoplasmic reticulum, where they initially assemble with cellular phosphatidylinositol (PtdIns) and/or PtdInsglycans, which are thought to stabilize the hydrophobic CD1d-binding pocket and facilitate CD1d transport to the cell surface [102-104]. Years' findings were further supported by studies demonstrating that recombinant microsomal triglyceride transfer protein (MTP) facilitates loading of immobilized CD1d with phosphatidylethanolamine [105-108]. After their arrival at cell surface, CD1d molecules engantigened late endosomal compartments by way of a tyrosine-based endocytic motif in their cytoplasmic tail [102]. Within late endosomal compartments, lipid transfers proteins, including saposins (Saps) and the GM2 activator (GM2A), edited antigens presented on CD1d molecules [109,110]. However, the contribution of individual lipid transfer proteins to CD1d-restricted T-cell responses remains poorly understood. A recent study has implicated apoE and the low-density lipoprotein receptor (LDLr) in the presentation of exogenous antigens by CD1d molecules [111], which were performed using a synthetic glycolipid, galactosyl (a1-2) galactosylceramide (GGC), which required lysosomal processing by carbohydrate hydrolases for conversion into the NKT cell antigenonist a-GalCer. The GGC capacity of activating iNKT cells was shown to be segregated with the very low density lipoprotein (VLDL) serum fraction, which included apoE. ApoE, which is associated with lipoprotein clearance and cellular cholesterol homeostasis, is required for efficient delivery of GGC to APCs and iNKT cell activation. Because LDLr-deficient cells were markedly defective in presenting GGC to iNKTcells, it was likely that apoE-bound glycolipids were taken up by APCs in an LDLr-dependent manner. It remained possible that additional lipoproteins and lipoprotein receptors played a role in CD1d-restricted antigen presentation. Taken together, these studies provided evidences for a remarkable connection between CD1d-resticted antigen presentation and the proteins that played a crucial role in lipid metabolism and atherogenesis. These findings were particularly intriguing in light of several reports documenting that NKT cells had a pro-atherogenic role [112-115].

2. Pathways of NKT Cells Promoting Asthma

Recently, two independent teams (Umetsu's group and our group) shed new light on NKT cells performance in asthma development (Fig. 2) [98, 99].

1) Recruitment

In asthma, our group discovered NKT cells (esp.$V\alpha24^+i$ NKT cells) selectively expressed CCR9 at high frequency (Fig. 2) [98]. When ligated with CCL25, activated CCR9-positive NKT cells were induced to migrate from their prevalent locations, i.e. the thymus, spleen, liver, and bone marrow [100], to penetrate into the blood vessels and then concentrate at the pathological focus in the airways [98]. Interestingly, CCR9/CCL25 ligation selectively induced the chemotaxis of NKT cells in patients with allergic asthma, but not normal NKT cells in healthy volunteers. CCR6 and CXCR3 were expressed on normal and asthmatic NKT cells with identical frequency. CCL20 (a ligand for CCR6) and CXCL9 (a ligand for CXCR3) did not induce different levels of chemotaxis activity between normal and asthmatic NKT cell.

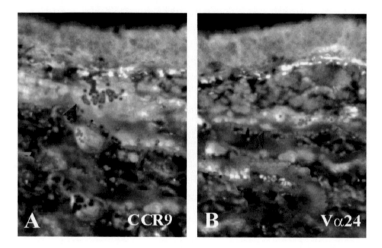

Figure 2. Morphology and distribution of $CCR9^+$ and $V\alpha24^+i$ NK T cells. The figure is showing the morphology and distribution of $CCR9^+$ (A) and $V\alpha24^+$ (B) iNKT cell in bronchi mucosa from the patients with allergic asthma detected by immunofluroescence confocal laser scanning microscopy with fluorescent ZnS/CdSe QD-conjugated antibody colloids. The mucosa-infiltrating iNKT cells expressing CCR9 or $V\alpha$ 24 chain were shown in bright blue as arrows indicated. Magnifications, ×400.

2) Indirect Pathway

Our group demonstrated an indirect pathway for NKT cells regulating the development of asthma. During their journey from the blood vessels to the airway bronchial mucosa, NKT cells might directly contact CD3+ T cells and polarize them to a Th2 bias [98].This regulatory function was dependent on DC participation and CCR9/CD226 synergic activation. CD226 is an adhesion molecule overexpressed on asthmatic NKT cells. CCL25/CCR9 ligation directly phosphorylate CD226, the lack of which will block the NKT cell--induced Th2-bias effect. This finding indicated that CCL25/CCR9 signals can cross-talk with CD226 signals to activate asthmatic NKT cells (Fig. 3, left panel).

3) Direct Pathway

At the same time, research from Umetsu's group confirmed the dominant roles of NKT cells located in the bronchial mucosa of asthma patients [99]. However, Umetsu showed a much different scenario, i.e. a direct pathway, in the final part of this process. Bronchial mucosa NKT cells expressed a large quantity of Th2 cytokines (IL-4, IL-13). In the mouse model, even without the help of conventional CD4$^+$ T cells and adaptive immunity, NKT cells can increase AHR, a distinctive feature of asthma (Fig. 3, right panel) [101]. It is clear that both direct and indirect pathways lead to the same effect, i.e. asthma antigengravation.

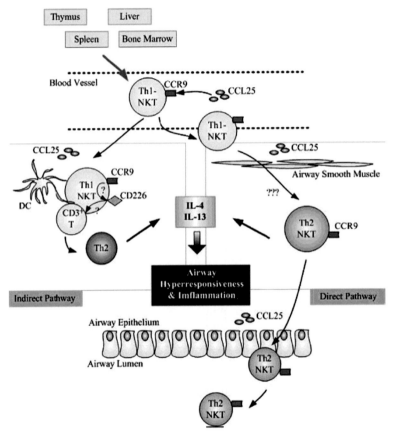

Figure 3. Pathways of NKT cells that promote asthma. Recruitment of asthmatic NKT cells in the airways - asthmatic NKT cells migrate from thymus, spleen, liver, and bone marrow, penetrate into the blood vessels, and then concentrate in the airway bronchial mucosa. Recruitment of NKT cells is dependent on the highly expressed CCR9 and igation of CCR9 with CCL25 (CCL25 is derived from dentritic cells, DCs, and airway tissue cells). During their journey, asthmatic NKT cells promote asthma in two different ways. Indirect pathway (left panel)-NKT cells directly contact CD3$^+$ T cells and polarize them to a Th2 bias, secreting large quantities of IL-4 and IL-13, which requires DC participation and the synergic signaling transduction between CCL25/CCR9 and CD226. Direct pathway (right panel)-while circulating in the blood, asthmatic NKT cells selectively highly express Th1 cytokines (IFN-γ). Once reaching the airway epithelium, asthmatic NKT cells shift to a Th2 bias, highly expressing IL-4 and IL-13, but not IFN-γ. Th2-biased NKT cells constitute most of the T cells in the airway bronchial mucosa and airway lumen. Both pathways lead to airway hyperresponsiveness and inflammation, i.e. asthma development.

3. Microorganism Infection and NKT Cell

Despite being a minority T-cell population, there is abundant evidence that NKT cells are important in mice for host defense against various bacteria, viruses and parasites. There are also data, especially for viral infections, that indicate this population could be important in human as well. As described below, last few years' work has begun to elucidate the two main mechanisms by which NKT cells help to protect host from infections.

1) Indirect NKT-Cell Activation

i. Cytokines and Endogenous Antigens Can Mediate NKT Cell Activation.

How do NKT cells with such limited TCR diversity respond to so many different infectious antigens? Recent studies have shed light on the mechanisms by which NKT cells are activated during infections (Fig. 4), including cases in which a microbial antigen for invariant TCR is not present. It was reported that human NKT cells produced IFN-γ in response to *Salmonella typhimurium* when cultured with DCs. *S. typhimurium* lipopolysaccharide (LPS) or recombinant IL-12 could induce IFN-γ production by NKT-cell clones co-cultured with DC, and IFN-γ production was inhibited by anti-IL-12 antibodies. Mouse DCs that are deficient for the adaptor myeloid differentiation primary-response gene 88 (MyD88), which interferes with much of the signalling by Toll-like receptors (TLRs), failed to induce IFN-γ. These data and results from *in vivo* experiments indicate that TLR engantigenement, mediated in this case by *Salmonella* sp. infection or by LPS, induces IL-12 synthesis that is crucial for the activation of NKT cells. This IL-12 was necessary but not sufficient, since the response could also be blocked with anti-CD1d antibodies [116].

More recently, it has been shown that DCs from *CD1d*[−/−] mice can not induce IFN-γ production by NKT cells in response to *S. typhimurium* [117]. Neither can mice deficient for β-hexosaminidase (HEXB), an enzyme that is required for the synthesis of isoglobo trihexosylceramide (iGb3) [118,119], the first candidate endogenous antigen. Because *S. typhimurium* does not contain lipid antigens for the NKT-cell TCR, these data indicate that NKT cells were activated by combination of IL-12, induced by LPS engantigenement of TLRs expressed by DCs, and the recognition of relatively weak self ligands, such as iGb3, presented by CD1d (Fig. 4a). In this response, TCR recognition of a foreign microbial antigen is not involved, and therefore we refer to this pathway as an 'indirect mechanism' for NKT cell activation (Fig. 4a–c) [120].

ii. Bacterial Infections Can Alter CD1d Expression.

Although recognition of endogenous antigens is required for NKT-cell activation in response to *S. typhimurium*, does bacterial infection increase CD1d expression and/or the presentation of stimulatory endogenous glycolipid ligands? Increased synthesis of the autologous GSL antigens including sulphatide, presented by CD1a or CD1b, and the ganglioside GM1, presented by CD1b, has been shown following stimulation of monocytes with bacteria or bacterial components [121]. Regarding the CD1d isoform, there are several reports of increased CD1d expression following exposure to bacteria or bacterial products. CD1d expression by DCs was increased when cultured *in vitro* with *S.typhimurium* or LPS from *Escherichia coli* [122], although the increased expression of CD1d on DCs was not observed *in vivo* during oral infection with *S. typhimurium*. The upregulation of CD1d was also observed on macrophages transferred into *Mycobacterium tuberculosis* infected mice. When mouse macrophages were cultured with *M. tuberculosis* bacteria, bacterial lipids or the synthetic TLR2 agonist Pam3Cys (tripalmitoyl-*S*-glyceryl cysteine), CD1d expression was increased, but only if recombinant IFN-γ was also added [123]. Recently, it was reported that *L. monocytogenes* induces increased CD1d expression on macrophages and DCs *in vitro* and *in vivo*, which could be inhibited by anti-IFN-β antibodies [124]. Macrophages with increased CD1d expression following exposure to bacteria or bacterial products were more effective at stimulating CD1d-dependent cytokine release by NKT cells, even in the absence of exogenous antigen. These data indicate that antigenpresenting cells (APCs) stimulated with bacteria or TLR antigenonists can induce the activation of NKT cells through the increased presentation of endogenous antigens by CD1d combined with IL-12 from activated DCs [125].

iii.   Endogenous Antigen-Mediated NKT Cells Activation.

Another mechanism of indirect NKT-cell activation was shown in a study of response to *Schistosoma mansoni* NKT cells were activated during *S. mansoni* infection and had an important role in the augmentation of the Th2 response [126,127]. When liver mononuclear cells (LMNCs) were cultured with *S. mansoni*-egg- sensitized DCs, the production of IFN-γ and IL-4 was significantly higher in cells from wild-type mice compared with cells from *Jα18*$^{-/-}$ mice or CD1d$^{-/-}$ mice. *S. mansoni* eggs do not contain glycolipid antigens for *i*NKT cell TCR, and DCs from Hexb$^{-/-}$ mice, unable to synthesize the endogenous antigen iGb3, failed to induce IFN-γ and IL-4 production by LMNCs [128]. These data indicate that *i*NKT cell activation by *S. mansoni*egg- sensitized DCs was mediated by the recognition of iGb3 [129]. However, it remains controversial whether iGb3 is in fact the sole or crucial endogenous ligand for NKT cells [130-135]. Indeed, that DCs from *Hexb*$^{-/-}$ mice are inhibited in their stimulation of NKT cells could be due to disruption of the endolysosomal vesicles where CD1d is loaded with antigen, as opposed to a deficiency in iGb3 synthesis[136].Surprisingly, DCs from *Il12*$^{-/-}$ mice or *MyD88*$^{-/-}$ mice could induce cytokine production by LMNCs in a similar manner to wild-type DCs. These data suggest that, in response to *S. mansoni* egg antigens, NKT cells can be activated by self glycolipids presented by CD1d, even in the absence of TLR signalling and IL-12 (Fig. 4b). It has not yet been determined, however, if *S. mansoni* egg extract increases the synthesis of endogenous glycolipid antigens and/or CD1d expression. Similarly, it is not known if *S. mansoni*-egg-sensitized DCs are activated in a TLR-independent manner to produce innate immune cytokines, other than IL-12, that might contribute to the activation of NKT cells [137].

iv.   Cytokine-Mediated NKT Cells Activation

It has recently been shown that NKT cells also can be activated by IL-12 and IL-18 produced by DCs activated by *E. coli* LPS, even in the absence of TCR stimulation by endogenous antigens presented by CD1d [138]. This purely cytokine-driven response constitutes a third type of indirect NKT-cell activation. When mice were injected with *E. coli* LPS, NKT cells produced IFN-γ but not IL-4, consistent with an inflammatory response that might enable NKT cells to contribute to the induction of Th1-type responses. Purified NKT cells produced IFN-γ in response to *E. coli* LPS when cultured with wild type DCs, but not with *Il12*$^{-/-}$ or *Il18*$^{-/-}$ DCs. Surprisingly, *CD1d*$^{-/-}$ DCs were also able to induce IFN-γ production by Vα14*i* NKT cells. Furthermore, physiological concentrations of IL-12 and IL-

18 together could induce IFN-γ production by purified NKT cells, even in the absence of DCs. These data indicate that inflammatory cytokines such as IL-12 and IL-18, which are produced by TLR-stimulated DCs and other APCs, are necessary and sufficient for induction of NKT-cell activation (Fig. 4c). Finally, a recent paper indicates a fourth indirect activation pathway, in which the cytokine and the TCR signals needed for NKT cell activation are delivered by two different DC subsets[139]. We conclude that NKT cells can be activated by infections indirectly through the activation of APCs even in the absence of microbial glycolipid antigens. By using any of the indirect mechanisms described above, NKT cells can sense the presence of many types of microorganism, including viruses that do not encode unique glycolipid antigens. This allows NKT cells to make a rapid response to a wide range of infectious antigens, similar to other cells that participate in the innate immune response. It is likely that a relatively weak TCR signal, delivered by self-glycolipid ligands presented by CD1d, can contribute to the signal delivered by innate-immune-cell cytokines. Furthermore, under some circumstances, such as those observed with exposure to *Schistosome* spp. egg extracts, TCR signal in reaction to self glycolipids presented by CD1d might be sufficiently potent to mediate NKT cell activation. Although indirect pathways are probably important, they do not account for the specificity and conservation of invariant TCR.

2) Direct Activation and Bacterial Antigens

i. Phospholipid Antigens

Several groups have searched for microbial glycolipid antigens that directly activate the NKT cells by engaging their invariant TCR. The first antigen reported was the glycophosphatidylinositol (GPI) anchor of proteins from *Plasmodium* spp. and *Trypanosoma* spp. It was asserted that the IgG responses to GPI-anchored surface antigens of *Plasmodium* spp. and *Trypanosoma* spp. were regulated through CD1d-restricted recognition of the GPI moiety by IL-4-producing NKT cells [140]. However, those findings remain controversial, as two subsequent studies did not reproduce the results [141,142]. Amprey *et al.* later showed that a subset of liver NKT cells could be activated by a lipophosphoglycan (LPG) extract from *L. donovani* [143]. *L. donovani* is part of genus *Leishmania*, the members of which are responsible for a wide array of diseases, ranging from self-healing cutaneous lesions to destructive skin and mucosal disease, and ultimately to fatal visceral infection. The authors observed an increase in parasite burden and a defect in the granulomatous response in *L. donovani*-infected mice deficient in NKT cells. They also showed that a small percentage (3-6%) of the NKT cells in the liver produced IFN-γ early after visceral *L. donovani* infection. They were able to show that this IFN-γ production by *i*NKT cells was IL-12 independent but CD1d dependent, indicating direct microbial antigen recognition by the NKT cell TCR instead of indirect recognition. *L. donovani* possesses a dense surface glycocalyx formed mainly by related glycoinositol phospholipids (GIPLs) and LPG. In a competitive assay, both purified GIPLs and LPG could inhibit α-GalCer-induced IL-2 production by Vα14*i* NKT cell hybridomas, indicating that GIPLs and LPG can bind to CD1d and be potential microbial antigens. However, LPG-CD1d tetramer staining of liver mononuclear cells could not be achieved, and injection of purified LPG could stimulate only 1.4% of the liver NKT cells *in vivo*. Therefore, these data indicate that LPG could activate only a subset of NKT cells. An early report on a potential microbial antigen for NKT cells showed that several mammalian lipids, namely phosphatidylglycerol (PG) and phosphatidylinositol (PI), could specifically bind to CD1d [144]. Moreover, PI, but not PG, could induce IFN-γ production by splenic T cells from Vα14-transgenic mice. Those results prompted the authors to see if the same were true for a mycobacterial variant of PI. They found that a subset of mouse Vα14*i* NKT cells and human Vα24*i* NKT cells reacted to a variant of PI, a purified glycolipid extracted from a mycobacterium cell-wall fraction and enriched for PI tetramannosides (PIM4) (Fig 4b). Analysis with PIM4-CD1d tetramers showed that the reactive cells comprised only a minority of the α-GalCer-CD1d tetramer-positive cells, especially in the mouse liver, indicating that PIM4 reactivity, similar to the *Leishmania* spp. LPG antigen, might be dependent on special features of the more variable β-chain of the semi-invariant TCR, as the α-chains are identical. Another limitation of this study was that only purified glycolipid extracted from mycobacteria was used, leaving open the possibility that a minor constituent of the purified product was responsible for the NKT-cell activation. Indeed, a later article reported that a synthetic version of PIM4 did not stimulate NKT cells, either *in vitro* or *in vivo* [145].

ii. Natural Glycosphingolipid Antigens from Sphingomonas Spp.

The studies on *Leishmania* spp- and mycobacteria-derived phospholipids helped establish the principle that the invariant TCR of NKT cells recognizes glycolipids from microbial antigens. However, because only a minority of the cells reacted, these specificities could not account for the conservation and selection of the invariant TCR. In 2005, the publication of three articles on GSLs from *Sphingomonas* spp. described antigens that are capable of activating essentially all NKT cells [117,146,147]. *Sphingomonas* spp. is Gram-negative bacteria that lack LPS and are highly abundant in the environment, including soil, sea water and plants. Prior to these studies, analyses of *Sphingomonas* spp. bacteria had revealed the presence of glycosylceramides in the cell wall, with structures similar to α-GalCer, including the rather unusual α-linked of the sugar to the sphingosine-containing lipid [148,149]. The structures of two of these are shown in Fig 4c. The abundance of these bacteria in the oceans indicates that they could have been in the original marine-sponge sample that was used to identify α-GalCer, and their structural similarity to this potent antigen made them good candidates for antigens able to activate the population of NKT cells. By using purified or synthetic versions of these GSLs in several different *in vitro* assays, it was shown that a galacturonic-acid-containing GSL (GalANTIGENSL), and one containing glucuronic acid (GlcANTIGENSL) (Fig 4c) and similar compounds [117,146] could bind to CD1d. Moreover, although not as potent as α-GalCer, the *Sphingomonas* spp. GSLs could specifically activate mouse NKT cells as well as human NKT cells. *Hexb*[−−] DCs also could stimulate NKT cells in response to *Sphingomonas* spp. These data indicate that the recognition of the putative self ligand iGb3 is not required for NKT cell activation in response to *Sphingomonas* spp [117]. Furthermore, TLR signaling and IL-12 were not required for GSL-mediated NKT cell activation, indicating that the observed activation was due to direct recognition of the microbial antigen by the TCR as opposed to the indirect pathway [117,146]. Direct recognition by the invariant TCR was confirmed by staining with CD1d tetramers loaded with the *Sphingomonas* spp. GSL. Indeed, in contrast to the LPG and PIM4 studies, CD1d tetramers loaded with synthetic GalANTIGENSL were able to detect at least 50% of the NKT cells, defined as cells reacting with α-GalCer-CD1d tetramers. Moreover, the reactive cells were absent in *CD1d*[−−] and *Jα18*[−−] mice, showing overlap between the α-GalCer-reactive and *Sphingomonas*-spp.-GSL-reactive populations [146]. *Ehrlichia* spp., which also belongs to the class of α-proteobacteria, can activate NKT cells independently of

TLR signaling and iGb3 synthesis as well; this indicates the direct recognition of microbial antigen, although this antigen has not yet been defined [117]. Mice deficient for NKT cells had a reduced clearance of *Sphingomonas* spp. bacteria at days 1-3 following infection [117,146]. The effect of NKT cells on clearance was most evident in liver [146], where these cells are most prevalent, but delayed clearance was also evident in lung [117]; eventually, however, the bacteria could be cleared even in mice lacking NKT cells. Therefore, direct recognition of GSLs by NKT cells can contribute to the early stages of host defense against LPS-negative microorganisms. However, when a high dose of *Sphingomonas* spp. bacteria was administered, mice with NKT cells died from shock whereas NKT cell-deficient mice were protected. This result provides another example in which an overactive protective NKT cell response is detrimental. The α-linked GSLs are believed to be unique to *Sphingomonas* spp. and related bacteria, but these bacteria are not highly virulent or pathogenic in humans, although infections in immunocompromised patients have been reported [150,151]. It remained uncertain whether NKT cells can recognize other classes of glycolipids that might be derived from pathogenic microorganisms.

iii. Galactosyl *Diacylglycerol Antigens* from Borrelia Burgdorferi

B. burgdorferi has recently been shown to have a different category of glycolipid antigen that activates TCR of NKT cells [141]. These spirochetes lack LPS, but contrary to *Sphingomonas* spp., they are pathogenic. *B. burgdorferi* is the causative antigen of Lyme disease, which, with 15,000 cases each year, is the most common vector-borne disease in the USA90. *CD1d*[-/-] mice infected with *B. burgdorferi* have been reported to have an increased bacterial burden, and they develop increased thickening of the tibiotarsal joint, indicative of arthritis. During *B. burgdorferi* infection, either after injection of live spirochetes or by using infected ticks, NKT cells were activated, as seen by an increase in both CD25 and CD69 expression on α-GalCer-CD1d tetramer-positive cells [141].

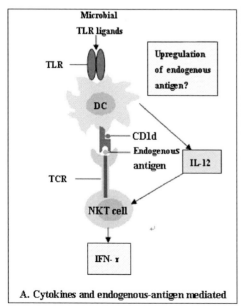
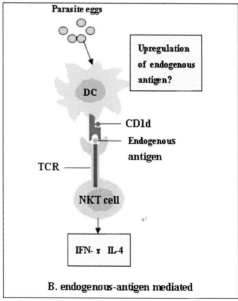

A. Cytokines and endogenous-antigen mediated   B. endogenous-antigen mediated

Figure 4. Continued on next page.

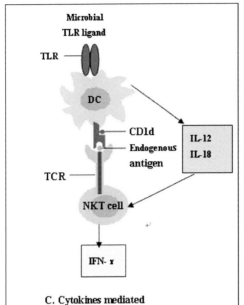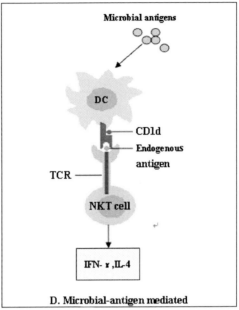

Figure 4. NKT cells have different pathways leading to their activation. a-c Indirect activation. The three indirect pathways do not depend on recognition by the NKT-cell T-cell receptor (TCR) of a microbial antigen, but depend on cytokine release by activated dendritic cells (DCs) and/or the recognition of endogenous glycolipid ligands. a Cytokine- and endogenous-antigen-mediated activation. During Salmonella typhimurium infection, lipopolysaccharide (LPS) stimulates Toll-like receptors (TLRs) on DCs and induces interleukin (IL)-12 release. NKT cells are activated by the combination of IL-12 produced by LPS-stimulated DCs and recognition of endogenous antigen presented by CD1d. It has not been determined if LPS induces upregulation of endogenous antigen in DCs. b Endogenous-antigen-mediated activation. Schistosoma mansoni egg-sensitized DCs induce IFN-γ and IL-4 production by NKT cells. In this response, TLR-mediated activation of DCs is not involved. However, recognition of endogenous antigen is required. It has not been determined if endogenous antigen is upregulated in S. mansoni egg-sensitized DCs. c Escherichia coli-LPS-stimulated DCs for IL-12 and IL-18 release. These cytokines are sufficient for IFN-γ production by NKT cells, and recognition of endogenous antigen presented by CD1d is not necessary for NKT-cell activation. d Microbial-antigen-mediated direct activation. Glycosphingolipids from Sphingomonas spp. and galactosyldiacylglycerols from Borrelia burgdorferi induce NKT-cell activation by engantigening their invariant TCRs. TLR-mediated DC activation, inflammatory cytokines such as IL-12 or recognition of endogenous antigen are not involved in this response.

Similar data were obtained when DCs were pulsed with bacterial lysates were injected into the mice. *B. burgdorferi* expresses two abundant glycolipids (*B. burgdorferi* glycolipid (BbGL)-I and BbGL-II) 91. BbGL-I is a cholesteryl 6-O-acyl-β-galactoside (structure not shown) and BbGL-II is a 1,2-diacyl- 3-O-α-galactosyl-*sn*-glycerol (Fig. 4d). BbGL-II, but not BbGL-I, was able to induce IL-2 secretion by Vα14*i* NKT cell hybridomas, although it was not as potent as α-GalCer or the *Sphingomonas* spp.-derived GalANTIGENSL. The BbGL-II isolated from the bacteria contained a mixture of C14:0, C16:0, C18:0, C18:1 and C18:2 fatty acids, with C16:0 and C18:1 being the most abundant, and it was uncertain which fatty acid(s) was linked to the *sn1* position of the glycerol and which was linked to the *sn2* position. Therefore, to further define NKT-cell specificity, eight chemically synthesized variants of BbGL-II were tested in several assays. BbGL-IIc, which consists of an oleic acid in the *sn1* position and a palmitic acid in the *sn2* position (Fig 4d) was by far the most potent antigen. By using MyD88-deficient mice or TRIF-deficient *Trif Lps2/Lps2* mice, the authors showed that BbGL-Iic induced *in vitro* proliferation, as well as *in vivo* activation of NKT cells independently of the presence of MyD88 or TRIF, indicating that NKT-cell activation induced by BbGL-IIc is not dependent on TLR signals and the indirect pathway, but rather is due to direct recognition of the antigen by the TCR of the NKT cells. This broad reactivity by the TCRs expressed by NKT cells was confirmed by tetramer staining. CD1d tetramers loaded with BbGLIIc detected approximately 23% of the liver NKT cells compared with α-GalCer-loaded tetramers. This might be an underestimate of the extent of reactivity to this relatively weak antigen in the NKT cell population, as BbGL-IIc was able to activate all of the NKT-cell hybridomas tested. Therefore, a substantial fraction of NKT cells can probably directly recognize α-galactosyl diacylglycerols derived from *B. burgdorferi* by direct engagement of their invariant TCR. The NKT-cell response to *B. burgdorferi* glycolipids is conserved in humans. Human NKT cell lines produced IFN-γ and IL-4 after culture with cells transfected with CD1d in the presence of the synthetic galactosyl diacyl glycerol compounds. The response pattern was different from mouse NKT cells; minimal cytokine release was induced by BbGL-IIc and maximal responses were observed after culture with compounds having a higher degree of unsaturation in the acyl chains, in particular BbGL-Iif (Fig4d). Because BbGL-IIc is a diacylglycerol-based molecule and not a GSL, these data have several important implications for understanding the biology of NKT cell responses to infectious antigens. First, because bacteria other than *B. burgdorferi* have glycoglycerol lipids, the invariant TCR could then have a broad reactivity to various microorganisms, which might in part explain the evolutionary selection for this TCR specificity. Second, by altering the degree of unsaturation of the acyl chains, we speculate that bacteria could evade recognition by NKT cells by producing glycolipids that can bind to CD1d but do not activate the invariant TCR.

4. NKT Cell-Mediated Liver Diseases

Studies from pre-clinical models have provided valuable insights on the functional role of NKT cells in the pathophysiology of liver diseases mediated by toxins, autoimmunity and viruses. From these studies, it is also clear that contribution of NKT cells to the pathophysiology of liver diseases is dependent on a number of factors: (i) the hepatotoxin mediating the liver injury; (ii) cytokines produced in the liver by these hepatotoxic antigens; and (iii) the ability of activated NKT cells to promote or suppress hepatic NK cell activation. The aforementioned studies also suggest that hepatic NKT cells can be activated either directly (Fig. 5a) or indirectly (Fig. 5b) by hepatotoxic antigens to exert beneficial or detrimental effects during liver diseases by selectively promoting the development of either pro-inflammatory or anti-inflammatory responses. Although this remains a contentious issue, it is apparent that activated hepatic NKT cells may contribute to liver injury predominantly via two mechanisms: (i) AICD or (ii) expansion/recruitment (Table 2). Given the mounting evidence, it is my contention that AICD of hepatic NKT cells may be a feature that is stimulus dependent and unique to liver diseases. In summary, the differences in the fate and function (i.e AICD versus expansion/recruitment) of hepatic NKT cells during liver diseases could potentially be attributed to the following: First and foremost, the quality of the stimulating signal which could affect TCR/ligand interaction and ultimately the function of NKT cells. Second, the efficacy of the CD1d expressing cell types in the presentation of antigen to hepatic NKT cells since a recent study reports that Kupffer cells and not dendritic cells are the key antigen presenting cell for hepatic NKT cells [165]. Third, the efficacy of NKT cells subsets may also potentially dictate the function of NKT cells during the liver diseases. For example, differences in the efficacy of the NKT cell subsets, CD4[+] T cell expressing NKT cells versus CD4[-] T cells expressing NKT cells, have been reported. Furthermore, not much is known about an NKT cell phenotype which does not express the invariant Vα14Jα18 TCR chain; interestingly, this NKT cell phenotype is exclusively expressed in the liver, is unreactive to α-GalCer and cannot be detected by the α-GalCer-CD1d tetramer [166].

Models of hepatic NKT call activation

Fig. 5a

Fig. 5b

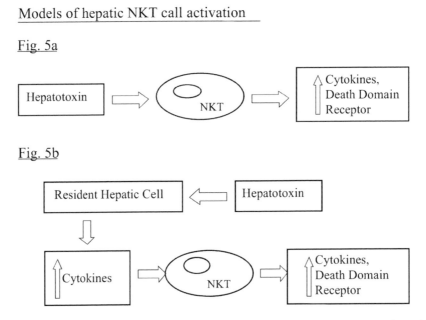

Figure 5a. In this model of direct hepatic NKT cell activation, the hepatotoxin may induce the liver to produce glycolipids, which provide ligands for CD1d on antigen-presenting cells, this in turn activate hepatic NKT cells resulting in increased hepatic expression of mediators (including cytokines and death domain receptors). These mediators could potentially directly promote or suppress hepatocyte cytotoxicity. Alternatively, the aforementioned mediators may modulate leukocyte recruitment into the liver, an effect that may subsequently induce or suppress hepatocyte cell death. Figure5b. This model of indirect hepatic NKT cell activation: (i) Resident hepatic cells (such as hepatocytes, Kupffer cells and/or endothelial cells) are activated by the hepatotoxic antigen; (ii) this results in increased hepatic production of cytokines; (iii) hepatic cytokines subsequently activate NKT cells in the liver; (iv) an effect which induces increased hepatic expression of mediators (including cytokines and death domain receptors). Next, these mediators may regulate the hepatic inflammatory response by promoting or suppressing hepatocyte cell death. On the other hand, the aforementioned mediators may modulate leukocyte recruitment into the liver, an effect that subsequently induce or suppress hepatocyte cytotoxicity.

Table 2.

Hepatotoxin	Effect of NKT cell deficiency and/or α-GalCer treatment on liver damage mediated by hepatotoxin	Fate of NKT cells	Contribution of NK cells to the effects of NKT cell
Toxin (Ethanol)	Reduced liver damaged in NKT cell KO mice; α-GalCer increased damage	Recruitment	Minimum
Toxin (AceTaminophen)	Reduced liver damaged in NKT cell KO mice	Recruitment	Significant
Autoimmune (Con A)	Reduced liver damaged in NKT cell KO mice	ACID	None
Autoimmune (α-GalCer)	α-GalCer treatment promotes liver damage in nomal mice and liver damage impaired in NKT cell KO mice	ACID & TCR downregulation	Minimum
Viral (mlCMV)	Viral replication in NKT cell KO mice similar to WT mice but liver damage was not determined	ACID	Unknown
Viral (mlCMV)	Viral replication in NKT cell KO mice similar to WT mice but replication was suppressed by α-GalCer. However, liver damage was not determined	Not determined	Significant
Viral (HBV transgenic)	Viral replication suppressed by α-GalCer but liver damage was not determined	? ACID	Significant

5. NKT Cell-Mediated Autoimmune Diseases

NKT cells are a unique lymphocyte subtype implicated in the regulation of autoimmunity, particularly diabetes and experimental allergic encephalomyelitis in animal models. Patients with diverse autoimmune diseases have reduced NKT-cell counts and, in diabetes and multiple sclerosis, effective NKT-cell regulation correlates with the secretion of Th2 cytokines [171]. Kojo S et al examined the reduction of TCR AV24$^+$, BV11$^+$CD4$^-$, CD8$^-$ (double-negative [DN]) NKT cells in peripheral blood lymphocytes from patients with rheumatoid arthritis, systemic lupus erythematosus, systemic sclerosis, and Sjogren's syndrome to analyze why NKT cells are selectively reduced in autoimmune diseases, and to examine whether nonresponse to α-GalCer is due to an abnormality in the antigen-presenting cells or NKT cells. They suggested that the reduced numbers of NKT cells in patients with autoimmune diseases may be due to an inadequate amount of α-GalCer-like natural ligands for the induction of NKT cells in vivo, or to a dysfunction in the NKT cells themselves [172].

1) Systemic Lupus Erythematosus

SLE is a systemic autoimmune disease characterized by inflammation in organs such as kidneys and presence of autoantibodies against nuclear antigens. Singh AK et al [173] demonstrated that NKT cell activation with α-GalCer suppresses or promotes pristane-induced lupus-like autoimmunity in mice, in a strain-dependent manner: Repeated in vivo treatment of pristane-injected BALB/c mice with the NKT cell ligand α-GalCer prior to the onset of florid disease suppressed proteinuria, in a manner dependent on CD1d and IL-4 expression. In sharp contrast, however, similar treatment of pristane-injected SJL mice with α-GalCer resulted in increased proteinuria. Consistent with these dichotomous effects of NKT cell activation on the development of lupus-like autoimmunity, NKT cells in BALB/c and SJL/J mice exhibited a mixed Th1/Th2 and a Th1-biased cytokine production profile, respectively. Green MR etal [207] found raised levels of IgG in patients with SLE and their relatives and high levels of IgG anti-dsDNA in patients were associated with low frequencies of NKT cells. These results suggest that NKT cells have an important role in the regulation of IgG production, although NKT cells with invariant T cell receptors may not necessarily be involved. NKT cells in the setting of SLE could lack the cytokine stimulus from NK or other cells that is needed to exert control on IgG production.

In our study, we report that NKT cells from active SLE patients are highly sensitive to anti-CD95-induced apoptosis compared with those from normal subjects and inactive SLE patients. CD226 expression is deficient on NKT cells from active SLE patients. The expression of one antiapoptotic member protein, survivin, is found to be selectively deficient in freshly isolated NKT cells from active SLE patients. CD226 preactivation significantly up-regulates survivin expression and activation, which can rescue active SLE NKT cells from anti-CD95-induced apoptosis. In transfected COS7 cells, we confirm that anti-CD95-mediated death signals are inhibited by activation of the CD226 pathway through stabilization of caspase-8 and caspase-3 and through activation of survivin. We therefore conclude that deficient expression of CD226 and survivin in NKT cells from active SLE is a molecular base of high sensitivity of the cells to anti-CD95-induced apoptosis. These observations offer a potential explanation for high apoptotic sensitivity of NKT cells from active SLE, and provide a new insight into the mechanism of reduction of NKT cell number in SLE and understanding the association between NKT cell deficiency and autoimmune diseases [174].

2) *Type I* Diabetes

NKT cells play a key regulatory role in type I diabetes. The absence of NKT cells correlates with exacerbation of type I diabetes, whereas an increased frequency and/or activation of NKT cells prevent β-cell autoimmunity. Various mechanisms are involved in the protective effect of NKT cells [175].

Multiple mechanisms have been proposed for the regulation of TID by NKT cells. In many cases, NKT cell-mediated protection against TID was associated with Th2 cytokines such as IL-4 and IL-10. Successful protection from TID mediated by adoptive transfer of thymic NKT cells required IL-4 and IL-10 [152], protection from diabetes in Vα14-Jα18 transgenic mice was associated with attenuated Th1 responses and enhanced Th2 responses against pancreatic islets [153], and protection mediated by α-GalCer was associated with Th2 deviation [154]. The possible role of Th2 deviation in disease protection is further supported by the finding that the α-GalCer analog C20:2, which was more effective in protecting NOD mice against diabetes than α-GalCer, promoted Th2 cytokine production by NKT cells [155]. However, OCH, another Th2 cytokine-biasing α-GalCer analog, had similar capacity as α-GalCer in disease protection [156]. One study showed that α-GalCer can protect IL-4- but not IL-10-deficient NOD mice against diabetes [157], but another study provided evidence that α-GalCer was able to protect NOD mice with a combined deficiency in IL-4 and IL-10 against diabetes [158]. In addition, one study investigating the capacity of NKT cells to protect NOD mice from disease induced by adoptive transfer of transgenic, diabetogenic CD4$^+$ T cells found evidence for a role of IFN-γ in disease protection [159]. A role for Th2 deviation was also suggested by some of the studies investigating NKT cells in human patients with TID, which have provided evidence for a Th1 bias in the cytokine profile of NKT cells from diabetics as compared with control subjects [160]. In concert, these findings suggest a role for Th2-dependent mechanisms in the regulatory role of NKT cells in TID (Fig. 6) but also reveal that Th2-independent mechanisms must be involved. NKT cells engage in intimate interactions with dendritic cells (DCs) and can influence the maturation and differentiation of DCs. As diabetes in NOD mice progresses, these animals develop an imbalance in the proportions of their DC subsets. It has been demonstrated that treatment of NOD mice with α-GalCer restores the frequency and functions of tolerogenic DCs in the pancreatic lymph nodes [161,162]. The α-GalCer analog C20:2 showed a similar and perhaps more pronounced correction of the DC imbalance in NOD mice [155]. Adoptive transfer of these α-GalCer-induced, tolerogenic DC to prediabetic NOD mice was able to protect animals against disease [162]. Recent studies have provided evidence for crosstalk between NKT cells and other subsets of regulatory T cells, in particular CD4$^+$CD25$^+$ regulatory T cells. One study has demonstrated that protection of TID mediated by adoptive transfer of α-GalCer-activated NKT cells requires the activity of CD4$^+$CD25$^+$ cells. Further, inactivation of CD25$^-$expressing cells before α-GalCer treatment abolished disease protection. These findings therefore suggest that protection from TID by α-GalCer-activated NKT cells requires the activity of CD4$^+$CD25$^+$ regulatory T cells. An additional mechanism that might contribute to the protective effects of NKT cells against TID is through functional inactivation of pathogenic T cells. Such a mechanism has been implicated in the capacity of NKT cells to suppress disease induced by adoptive transfer of transgenic, diabetogenic CD4$^+$ T cells. NKT cells impaired the differentiation of these islet-specific T cells into Th1 effectors and instead induced unresponsiveness or anergy in these cells. The mechanisms by which NKT cells induced such anergy in pathogenic T cells remain unclear but CD1d expression did not appear to be required, despite the need for cellular contact. Collectively, these findings provide evidence that NKT cells regulate TID by both Th2-dependent and Th2-independent mechanisms (Fig. 6). Some of these proposed mechanisms may be linked to each other. For example, activated NKT cells might promote the development of tolerogenic DCs, which, in turn, might induce the generation of CD4$^+$CD25$^+$ regulatory T cells that suppress pathogenic Th1 and CD8$^+$ cytotoxic T cell responses against islet autoantigens [163, 164].

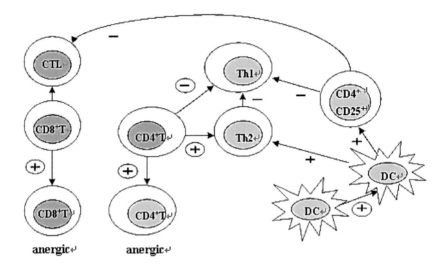

Figure 6. Possible mechanisms involved in the regulation of TID by NKT cells.

NKT cells promote Th2 responses, induce anergy in autoantigen-specific T cells, induce the differentiation of dendritic cells (DC) into tolerogenic DC, and promote the activities of CD4$^+$CD25$^+$ regulatory T cells. In concert, these mechanisms might dampen the activities of pathogenic Th1 cells and cytotoxic T lymphocytes (CTL). Circled symbols identify steps where NKT cells are thought to directly impact immunoregulatory mechanism.

6. Mechanisms of NKT Cell-Mediated Transplant Tolerance

The mechanism by which CD1d-restricted Vα14 NKT cells participate in transplant tolerance has yet to be completely clarified. Recent study showed that repeated activation of NKT cells by their specific glycolipid ligand, α-GalCer, leading to a change in function to an immune regulatory role with IL-10 production. Moreover, these cells were shown to be able to induce regulatory DCs. In Jiang etals' study, they showed that NKT cells from transplant tolerant recipients of cardiac allograft produced higher levels of IL-10, which is required for the maintenance of tolerance; this was proved by adoptive transfer experiments. In addition, DCs from wildtype (WT) tolerant recipients but not NKT cell-deficient recipients showed a higher IL-10-producing profile, a more immature phenotype, and tolerogenic capability. CD4 T cells from WT tolerant recipients but not NKT cell-deficient recipients also produced higher levels of IL-10 upon alloantigen stimulation and showed lower proliferative activity that was reversed by blocking the IL-10 receptor. These data indicate the existence of IL-10-dependent immune regulatory interplay among NKT cells, DCs, and CD4 T cells, even in the absence of artificial stimulation of NKT cells with synthetic glicolipids, which is required for the maintenance of transplant tolerance [176].

7. Differential Antitumor Immunity Mediated by NKT Cell Subsets

The existence of functionally distinct NKT cell subsets in vivo may shed light on the long-appreciated paradox that NKT cells function as immunosuppressive cells in some disease models, whereas they promote cell-mediated immunity in others [167,168]. Nadine Y. Crowe et al [169] showed previously that NKT cell-deficient TCR Jα18$^{-/-}$mice are more susceptible to methylcholanthrene (MCA)-induced sarcomas, and that normal tumor surveillance can be restored by adoptive transfer of wild type liver-derived NKT cells. Furthermore, when CD4^{+} and CD4^{-}liver-derived NKT cells were administered separately, MCA-1 rejection was mediated primarily by the CD4 fraction. Very similar results were achieved using the B16F10 melanoma metastasis model, which requires NKT cell stimulation with α-galactosylceramide. The impaired ability of thymus-derived NKT cells was due, in part, to their production of IL-4, because tumor immunity was clearly enhanced after transfer of IL-4-deficient thymus-derived NKT cells. Masaki Terabe et al [170] also showed that type I NKT cells were not necessary for the immune suppression. Type II NKT cells may be sufficient for negative regulation, whereas protection has been found to be mediated by α-galactosylceramide-responsive type I NKT cells.

V. NKT Cells-Associated Therapy

Recent understanding of NKT cells performance in the development of various diseases might unveil new therapy targets and management strategies. Our observations offer a potential explanation for high apoptotic sensitivity of NKT cells from active asthma and SLE. Nowadays, there is continued enthusiasm for the development of NKT cell-based therapies of human diseases.

1. Production and Characterization of Monoclonal Antibodies

Recent discovery that structural analogues of α-GalCer such as OCH or C20:2 can elicit different NKT cell responses than the strong agonist KRN7000 provides a practical approach to the generation of therapeutic immunomodulators for a range of different applications. The α-GalCer known as KRN7000 remains the best studied ligand of the lipid-binding MHC class I-like protein CD1d. The KRN7000:CD1d complex is highly recognized by NKT cells, an evolutionarily conserved subset of T lymphocytes that express an unusual semi-invariant T cell antigen receptor, and mediate a variety of proinflammatory and immunoregulatory functions. To facilitate the study of glycolipid antigen presentation to NKT cells by CD1d, someone undertook the production of mouse monoclonal antibodies (mAbs) specific for complexes of KRN7000 bound to mouse CD1d (mCD1d) proteins. They found three such monoclonal antibodies were isolated that bound only to mCD1d proteins that were loaded with KRN7000 or closely-related forms of α-GalCer. These mAbs showed no reactivity with mCD1d proteins that were not loaded with α-GalCer, nor did they bind to complexes formed by loading mCD1d with the self-glycolipid and putative NKT cell ligand isoglobotrihexosylceramide. These complex-specific monoclonal antibodies allow the direct detection and monitoring of complexes formed by the binding of KRN7000 and other α-GalCer analogues to mCD1d.Therefore, the availability of these mAbs should facilitate a wide range of studies on the biology and potential clinical applications of CD1d-restricted iNKT cells [177].

2. Expansion of NKT Cells From Cord Blood Mononuclear Cells Using IL-15, IL-7 and Flt3-L Depends on Monocytes

Human NKT cells are a unique T cell population specifically and potently activated by α-GalCer presented by CD1d. so Hikaru Okada etal present a simple and efficient method for expanding NKT cells from human cord blood mononuclear cells (CBMNC) using α-GalCer in the presence of interleukin (IL)-15, IL-7 and Flt3-L. The addition of α-GalCer from day 0, compared to its addition from day 8 or day 15, induced a greater expansion of NKT cells. The maximal expansion of NKT cells was observed after 15 days (2300-fold). Thereafter, the number of NKT cells decreased slowly, a decrease that was correlated with the diminution of CD1d-positive cells. NKT cell proliferation induced by α-GalCer was not observed when CD1d-expressing monocytes were depleted from CBMNC, whereas B cell and dendritic cell depletions had no effect. Expanded NKT cells were $CD4^+CD8^-$ and secreted both IL-4 and IFN-γ. In this system, $CD3^+$ T cells and $CD3^-CD56^+$ NK cells were also expanded. However, the expansion of NKT cells had no significant functional effect on T and NK cells. This expansion method of CBMNC-derived NKT cells is simple and may be helpful for clinical use [178].

3. NKT Cell-Based Therapies in Several Diseases

1) Asthma

Several studies indicated airway hyperreactivity (AHR), eosinophilic inflammation with a Th2-type cytokine profile, and specific Th2-mediated IgE production characterize allergic asthma [179]. Indeed, upon activation by glycolipid antigens, they rapidly secrete both Th1 and Th2 cytokines, which affects the development of later immune responses. In mouse models, someone treatmented with α-GalCer, a specific NKT ligand, the method inhibits most of the parameters associated with allergic asthma. Increased IFN-γ synthesis, rather than regulatory IL-10, accounts for this paradoxical effect [180]. Hachem P etals' findings demonstrated for the first time that α-GalCer administered locally inhibits asthma symptoms, even in predisposed asthmatic mice, through an NKT cell- and IFN-γ-dependent pathway [181]. Ikegami Y etal showed the NKT cell counts of asthmatic patients were significantly lower than those of healthy controls [182]. Therefore, recent studies suggest that therapies targeted at depletion or limiting of NKT cells may be a possible strategy for the treatment of asthma.

2) Inflammatory Bowel Diseases (IBD)

Therapeutic strategies that are designed to activate specifically Vα14 NKT cells may prove to be beneficial in treating intestinal inflammation. Administration of a single dose of OCH attenuated colonic inflammation, as defined by body weights and histologic injury. The protective effects of OCH could not be observed in Vα14 NKT cell-deficient mice. In vivo treatment with OCH had improved the IFN-γ/IL-4 ratio from colonic LPLs, indicated that the activation of Vα14 NKT cells by OCH plays a pivotal role in mediating intestinal inflammation via altered mucosal T-helper type I/type 2 responses [198]. Induction of oral immune regulation significantly ameliorated disease activity. Immune regulation via oral administration of CEP is a safe and possibly effective treatment for subjects with moderate CD and may provide means of antigen-specific immune modulation [197]. In Krajina T et al experiment, cLP DC might be interesting targets for novel therapeutic approaches to modulate mucosal T cell responses in situ. They reported CD11c$^+$ (F4/80(-) CD68(-)) dendritic cells in the colonic lamina propria (cLP) of normal and immunodeficient (RANTIGEN1($^{-/-}$)) C57BL/6 (B6) mice show high surface expression of MHC class I/II molecules and CD1d, and low surface expression of CD40, CD80, CD86 costimulator molecules. CD4$^+$ alpha β T cells from normal or MHC class II-deficient B6 mice transferred into congenic RANTIGEN1($^{-/-}$) hosts induce a progressive, lethal colitis. [199].

3) Hepatitis

Glucocerebroside (GC) is a naturally occurring glycolipid. The beneficial effect of GC was associated with a 20% decrease of intrahepatic NKT lymphocytes, significant lowering of serum IFN-γ levels, and decreased STAT1 and STAT6 expression. In vitro administration of GC led to a 42% decrease of NKT cell proliferation in the presence of dendritic cells but not in their absence. Intraperitoneally administered radioactive GC was detected in the liver and bowel. So administration of GC led to amelioration of ConA hepatitis associated with an inhibitory effect on NKT lymphocytes and GC holds promise as a new immune-modulatory antigen [196].

4) Autoimmune Diseases

NKT cells as targets for immunotherapy of autoimmune diseases, and the identification of α-GalCer as a potent stimulator of NKT cells led many laboratories to investigate the effects of NKT cell activation on the regulation of immune responses [183]. Someone revealed that α-GalCer induces rapid and robust cytokine production by NKT cells, secondary activation of a variety of innate and adaptive immune cells, and modulation of Th cell responses. All these studies have raised significant enthusiasm for manipulation of NKT cells as a means of preventing autoimmunity in the clinical setting, so there are significant concerns regarding the safety of repeated α-GalCer injections in human subjects [64,184]. In addition☐the unique glycolipid-reactive lymphocytes are known to produce a large quantity of Th2 cytokines such as IL-4 when encountering their ligands like α-GalCer. Whereas most of the NKT ligands so far described would stimulate both Th1 and Th2 cytokine production by NKT cells, the synthetic compound OCH, an analogue of α-GalCer with a shorter lipid tail, is the one which selectively induces IL-4 production. Given this property, oral or intraperitoneal OCH administration would prohibit the development of a variety of Th1-mediated pathology, including EAE, collagen-induced arthritis, type I diabetes, DSS-induced colitis and acute GVHD by inducing Th2 bias [185].

i. SLE

NKT cells have an important role in the regulation of IgG production, although NKT cells with invariant T cell receptors may not necessarily be involved. In the setting of SLE NKT cells could lack the cytokine stimulus from NK or other cells that is needed to exert control on IgG production. Through the pathway to enhance NKT cell activity, may provide a novel basis for therapy in SLE [187]. In addition, Nemoto K etal reported Deoxyspergualin (DSG) was found to be a promising antigen for curing such established autoimmune disease [188].

ii. Type I Diabetes

NKT cells can prevent diabetes by inhibiting the differentiation of anti-islet T cells. Multiple-dose α-GalCer treatment regimen, which is known to promote a dominant Th2 environment, can prevent onset of spontaneous and cyclophosphamide (CY)-accelerated TID. This protection is associated with elevated IL-4 and IL-10 in the spleen and pancreas of protected female NOD mice. Concomitantly, IFN-γ levels are reduced in both tissues. More importantly, the protective effect of α-GalCer in CY-accelerated TID is abrogated by the in vivo blockade of IL-10 activity. So, α-GalCer treatment significantly prolongs syngeneic islet graft survival in recipient diabetic NOD mice [186].

5) Transplant Tolerence

NKT cells play crucial roles in preventing transplantation tolerance. Cyclosporin A (CsA) generally used in clinical transplantation therapy, could modulate NKT cell activation induced by α-GalCer treatment. CsA suppresses α-GalCer-induced cytokine productions and dendritic cell maturation of mouse NKT cells [189].

In addition, Li W etal found an excellent in vivo model to generate regulatory T cells to allospecific transplant antigens, which is an experimental colitis model, show that a subtherapeutic dose of FK506 enhanced the tolerating effect of NKT cells induced by oral tolerance, prolonging allograft survival by generating CD25$^+$CD4$^+$ CTLA4 T cells. [190].

Graft-versus-host disease (GVHD) and graft-versus leukemia (GVL) effects are closely related to each other after allogeneic stem cell transplantation. NKT cells act as a double-edged sword in the progressive or suppressive effects on diseases. Such contradictory phenomena may be related to the function or types of APCs in response to their ligand. A single-dose injection of α-GalCer can induce immunity through fully mature dendritic cells in an antigen-specific manner. By contrast, multiple injections of α-GalCer would induce tolerance, which may be caused by immature APCs. This response suggests that the function of NKT cells can be determined by α-GalCer for controlling the immune response. Furthermore, activation of NKT cells followed by activation of APCs and IL-12 production may lead to activation of NK cells and suppress GVHD in mismatched major histocompatibility complex combinations or may induce GVL effects. Control and modification of NKT cell function may play an important role in regulating GVHD/GVL effects [191]. Moreover, to evaluate the effect of transplantation of NKT lymphocytes on graft versus host disease (GVHD) in a murine model of semiallogeneic BMT. Tolerance induction was associated with an increased peripheral CD4/CD8 ratio, intrahepatic trapping of CD8$^+$ lymphocytes and a shift towards a Th2 type cytokine profile, manifested by decreased IL-12/IL-10, IL-12/IL-4, IFN-γ/IL-10, and IFN-γ/IL-4 ratios. Accordingly, transplantation of DX5$^+$ cells holds promise as a novel therapeutic measure for GVHD [192].

In corneal allograft transplantation, BALB/c mice that tolerate the allogeneic grafts develop allogeneic-specific anterior chamber-associated immune deviation. Also NKT cells are required for induction of allospecific Tr cells and are essential for survival of corneal allografts, Thus mechanisms that contribute to cornea graft acceptance may lead to new therapies for improvement in graft survival in high-risk corneas and other transplanted tissues and grafts [193].

Pancreatic islet transplantation represents a potential treatment for insulin-dependent diabetes mellitus. Ikehara Y etal indicated that NKT cells play a crucial role in the acceptance of rat islet xenografts in mice treated with anti-CD4 antibody, probably by serving as immunosuppressive regulatory cells. In their experiment, an anti-CD4 mAb, administrated after transplantation, allowed islet xenografts to be accepted by C57BL/6 mice, with no need for immunosuppressive drugs [194].

Permanent acceptance of donor skin graft is readily induced in the MHC-matched and minor Antigen-mismatched recipients after treatment with donor spleen cells and cyclophosphamide (CP). Iwai T etal suggested that NKT cells are essential for CP-induced tolerance and may have a role in the establishment of mixed chimerism, resulting in clonal deletion of donor-reactive T cells in the recipient thymus [195].

6) HIV

A critical component of the host's innate immune response involves lipid antigen presentation by CD1d molecules to NKT cells. The reciprocal regulation of CD1d-mediated Antigen presentation by MAPK suggests that the targeting of these pathways is a novel means of immune evasion by viruses [200]. Markus Moll et al demonstrated innate NKT cells are infected and lost in HIV-1-infected patients, and this could contribute to HIV-1 pathogenesis, because NKT cells play an important role in directing both adaptive and innate immunity. They showed administration of IL-2 to HIV-1-infected patients leaded to substantial and sustained $CD4^+$ T cell expansion, involving both naive and memory cells, also indicated that IL-2 treatment in combination with effective ART is beneficial for the restoration of innate NKT cell immunity in patients with primary HIV-1 infection [201]. A study has demonstrated that a bispecific lentiviral vector could be used to stably deliver shRNAs targeted to both CCR5 and CXCR4 coreceptors into $CD34^+$ hematopoietic progenitor cells and derive transgenic macrophages. Transgenic macrophages with down regulated coreceptors were resistant to both R5 and X4 tropic HIV-1 infections. It is now possible to construct gene therapeutic lentiviral vectors incorporating multiple shRNAs targeted to cellular molecules that aid in HIV-1 infection. Use of these vectors in a stem cell setting shows great promise for sustained HIV/AIDS gene therapy [206].

VI. Further Investigation of NKT Cells in Disease

NKT cells express markers of both NK cells and T lymphocytes. These include Ly49 family receptors, NK1.1 and TCR. A hallmark of NKT cells is copious IL-4 and IFN-g secretion promptly upon TCR activation [202]. Following TCR expression, CD1d restricted cells can be identified even before NK1.1 expression based on the ability to bind α-GalCer/CD1d tetramers [203]. So scientists leave a question to us whether NKT cells represent a distinct lineage or are a subset of the mainstream T cell pool?

The best evidence that NKT cells are derived from DP thymocytes comes from studies in which the intrathymic injection of purified TCR⁻CD4⁺CD8⁺ thymocytes into genetically marked recipients gave rise to NKT cells. NKT cells can be further subdivided on the basis of the cell surface marker DX5 expression into four distinct populations of CD1d-tetramer-reactive cells. Stage 1: DX5⁻NK1.1⁻ (most immature); Stage 2: DX5⁺NK1.1⁻; Stage 3: DX5⁺NK1.1⁺; and, Stage 4: DX5⁻NK1.1⁺. Stage three and four represent the most mature cells, and are present in approximately equal numbers in the Adult. Thus To definite the intermediate Stage of NKT-cell development, more work is needed to identify earlier progenitors and further clarify their lineage relationship to mainstream T cells. In comparison to mature NK1.1⁺ cells, NK1.1⁻ cells are larger, exhibiting a naive phenotype: the TCR density is higher than on mature NKT cells, expressing a heterogeneous low level of CD44 and very little CD69 and making little IFN-γ but producing larger amounts of IL-4 than mature cells, but NK1.1⁺ is converse. There is some transcriptional machinery regulating NKT cell development. Ets-1 appears to be selectively required during NKT-cell ontogeny, as does AP-1 activity. NF-κB family members are also needed. The most profound effect is observed in relB mutants. This mutation operates in a cell non-autonomous manner, suggesting that the stroma secretes a factor that regulates NKT-cell development or expansion. One candidate for this factor is IL-15, as it is decreased in the stroma of relB mutants and is important in NKT cell development [202]. To explain the development of Vα14 iNKT cells, two models are established. In the "instructional" model, expression of Vα14 iTCR instructs thymocytes to become (Vα14i) NKT cells, perhaps because of a high-affinity interaction with CD1d-presented ligands or interaction with double-positive thymocytes. According to the precommitment model, a precursor decides to become a (Vα14i) NKT cell before TCR rearrangement [204] that means a distinct pre-NKT-cell pool exists in the thymus. So far, most evidence favors the 'instructive' model, including the observation that DP thymocytes can give rise to NKT cells [203].

Correlations between disease status and NKT cell function have been reported in several cases [204]. NKT cells play a critical role in infectious disease processes. The mechanism by which NKT cells exert their protective effects against viruses has not yet been characterized and evidence for the role of NKT cells in both bacterial and fungal infections is limited. But it has been shown that the rapid cytokine production by activated NKT cells induces bystander activation of NK, T and dendritic cells, thus, favoring development of T cell responses responsible for limiting the viral infection. NKT cells appear to be involved in immunity to a number of viruses including herpes simplex viruses (HSV) and hepatitis B. This required the expression of IFN-γ and TNF-α/β, and both of these genes are detected at high levels in the livers of mice injected with α-GalCer [203]. A number of investigators have reported decreased NKT cells in a variety of autoimmune diseases [205]. Such as diabetes, both murine models of type I diabetes (NOD, non-obese diabetic mice) and human diabetes are prevented by NKT cell activation ;while others, such as atherosclerosis, are worsened by NKT cell activation. The size of atherosclerotic lesions in mice is enhanced by administration of α-GalCer, dependent on CD1d or Jα18 expression. This is most likely due to enhanced expression of proinflammatory cytokines by NKT cells, leading to enhanced activation of macrophages, B and T cells that are attracted to the aorta [203]. Multiple sclerosis (MS) is a chronic inflammatory disease of the central nervous system. A decrease in Vα24i mRNA in the peripheral blood of MS patients has been found, with subsequent work suggesting that this decrease was confined to the CD4+ subset of T cells. Furthermore, short-term cell lines of

NKT cells from MS patients in remission had a Th2 cytokine bias compared to similar lines from relapsed patients, suggesting a regulatory role for NKT cells [205]. There is a potent role for α-GalCer in preventing tumor metastases, but in the absence of α-GalCer treatment, tumor surveillance is mainly unaffected when (Vα14i) NKT cells are absent, although the response to methylcholanthrene-induced sarcomas and GM-CSF-transfected tumors constitute exceptions in which (Vα14i) NKT cells play an important role even in the absence of α-GalCer. In advanced prostate cancer patients, for example, expanded lines of NKT cells had a reduced capacity to produce IFN-γ upon stimulation. Although intravenous administration of α-GalCer had only limited effects in cancer patients, injection of DCs pulsed with α-GalCer did activate innate and acquired immunity, including increases in serum IFN-γ . Therefore, as evidenced in mice, transferring DCs pulsed with α-GalCer could lead to a much more potent Th1 response than injecting the free compound [204]. NKT cell numbers are reduced in the peripheral blood of patients with a variety of organ-specific and systemic autoimmune conditions. Beneficial NKT cell reactivity to tumors may provide an example of an anti-self response that is harmful in the context of autoimmune disease, and an overly robust Th2 response by these cells can be detrimental [205].

It can be seen NKT cells are now recognized as bridging the innate and adaptive immune systems. Although important progress has been made regarding the specificity and effector functions of these cells, much remains to be learned. Important areas of future research include: the role of iGb3 and other endogenous glycolipids in NKT cell development and function; the specificity and function of distinct lipid-binding proteins in CD1d-restricted antigen presentation; the relevance of the relationship between CD1d-restricted antigen presentation and lipid metabolism to the role of NKT cells in atherosclerosis; the correlation of NKT cell effector functions with the diverse immunomodulatory roles of these cells; the elucidation of the immune functions of distinct NKT cell subsets and subtypes; improved understanding of the similarities and differences between NKT cells in mice and humans; and the development of rational NKT cell based immunotherapies by taking advantage of NKT cell agonists that promote distinct NKT cell effector functions, delivery of NKT cell agonists in the context of APCs or in combination with other immunotherapeutics, and targeting distinct NKT cell subsets or subtypes [208].

References

[1]   Sandberg, JK; Ljunggren, HG. Development and function of CD1d-restricted NKT cells: influence of sphingolipids, SAP and sex. *Trends Immunol,* 2005, 26, 347-349.

[2]   Godfrey, DI; MacDonald, HR; Kronenberg, M; Smyth, MJ; Van Kaer, L. NKT cells: what's in a name? *Nat. Rev. Immunol,* 2004, 4, 231-237.

[3]   Makino, Y; Kanno, R; Koseki, H; Taniguchi, M. Development of Valpha4+ NK T cells in the early stages of embryogenesis. *Proc. Natl. Acad. Sci. U. S. A.,* 1996, 93, 6516-6520.

[4]   Sato, H; Nakayama, T; Tanaka, Y; Yamashita, M; Shibata, Y; Kondo, E; Saito, Y; Taniguchi, M. Induction of differentiation of pre-NKT cells to mature Valpha14 NKT cells by granulocyte/macrophage colony-stimulating factor. *Proc. Natl. Acad. Sci. U. S. A,* 1999, 96, 7439-7444.

[5]   Maeda, M; Shadeo, A; MacFadyen, AM; Takei, F. CD1d-independent NKT cells in beta 2-microglobulin-deficient mice have hybrid phenotype and function of NK and T cells. *J. Immunol,* 2004, 172, 6115-6122.

[6]   Bendelac, A; Killeen, N; Littman, DR; Schwartz, RH. A subset of CD4+ thymocytes selected by MHC class I molecules. *Science,* 1994, 263, 1774-1778.

[7]   Mendiratta, SK; Martin, WD; Hong, S; Boesteanu, A; Joyce, S; Van Kaer, L. CD1d1 mutant mice are deficient in natural T cells that promptly produce IL-4. *Immunity,* 1997, 6, 469-477.

[8]   Xu, H; Chun, T; Colmone, A; Nguyen, H; & Wang, C, R. Expression of CD1d under the control of a MHC class Ia promoter skews the development of NKT cells, but not CD8+ T cells. *J. Immunol.* 2003, 171, 4105–4112.

[9]   Benlagha, K; Weiss, A; Beavis, A; Teyton, L; Bendelac, A. In vivo identification of glycolipid antigen-specific T cells using fluorescent CD1d tetramers. *J. Exp. Med,* 2000, 191, 1895-1903.

[10]   Godfrey, DI; Berzins, SP. Control points in NKT-cell development. *Nat. Rev. Immunol,* 2007, 7, 505-518.

[11]   Hammond, KJ; Pelikan, SB; Crowe, NY; Randle-Barrett, E; Nakayama, T; Taniguchi, M; Smyth, MJ; van, Driel, IR; Scollay, R; Baxter, AG; Godfrey, DI. NKT cells are phenotypically and functionally diverse. *Eur. J. Immunol,* 1999, 29, 3768-3781.

[12]   Seino, K; Taniguchi, M. Functionally distinct NKT cell subsets and subtypes. *J. Exp. Med,* 2005, 202, 1623-1626.

[13]   Pellicci, DG; Hammond, KJ; Uldrich, AP; Baxter, AG; Smyth, MJ; Godfrey, DI. A natural killer T (NKT) cell developmental pathway iInvolving a thymus-dependent NK1.1 (-)CD4(+) CD1d-dependent precursor stage. *J. Exp. Med,* 2002, 195, 835-844.

[14]   Egawa, T; *et al.* Genetic evidence supporting selection of the Vα14iNKT cell lineage from double-positive thymocyte precursors. *Immunity* 2005, 22, 705–716.

[15]   Levitsky, HI; Golumbek, PT; Pardoll, DM. The fate of CD4-8- T cell receptor-alpha beta+ thymocytes. *J. Immunol,* 1991, 146, 1113-1117.

[16]   Bezbradica, J, S; Hill, T; Stanic, A. K; Van, Kaer, L; & Joyce, S.Commitment toward the natural T (iNKT) cell lineage occurs at the CD4$^+$8$^+$ stage of thymic ontogeny. *Proc. Natl. Acad. Sci. U. S. A,* 2005, 102, 5114–5119.

[17]   Gapin, L; Matsuda, JL; Surh, CD; Kronenberg, M. NKT cells derive from double-positive thymocytes that are positively selected by CD1d. *Nat. Immunol,* 2001, 2, 971-978.

[18]   Egawa, T; Eberl, G; Taniuchi, I; Benlagha, K; Geissmann, F; Hennighausen, L; Bendelac, A; Littman, DR. Genetic evidence supporting selection of the Valpha14i NKT cell lineage from double-positive thymocyte precursors. *Immunity,* 2005, 22, 705-716.

[19]   Bezbradica, JS; Hill, T; Stanic, AK; Van, Kaer, L; Joyce, S. Commitment toward the natural T (iNKT) cell lineage occurs at the CD4+8+ stage of thymic ontogeny. *Proc. Natl. Acad. Sci. U. S. A,* 2005, 102, 5114-5119.

[20]   Berzins, SP; McNab, FW; Jones, CM; Smyth, MJ; Godfrey, DI. Long-term retention of mature NK1.1+ NKT cells in the thymus. *J. Immunol,* 2006, 176, 4059-4065.

[21]   Borowski, C; Bendelac, A. Signaling for NKT cell development: the SAP-FynT connection. *J. Exp. Med,* 2005, 201, 833-836.

[22]   Kunisaki, Y; Tanaka, Y; Sanui, T; Inayoshi, A; Noda, M; Nakayama, T; Harada, M; Taniguchi, M; Sasazuki, T; Fukui, Y. DOCK2 is required in T cell precursors for development of Valpha14 NK T cells. *J. Immunol*, 2006, 176, 4640-4645.

[23]   Elewaut, D; Lawton, AP; Nagarajan, NA; Maverakis, E; Khurana, A; Honing, S; Benedict, CA; Sercarz, E; Bakke, O; Kronenberg, M; Prigozy, TI. The adaptor protein AP-3 is required for CD1d-mediated antigen presentation of glycosphingolipids and development of Valpha14i NKT cells. *J. Exp. Med*, 2003, 198, 1133-1146.

[24]   Kojo, S; Seino, K; Harada, M; Watarai, H; Wakao, H; Uchida, T; Nakayama, T; Taniguchi, M. Induction of regulatory properties in dendritic cells by Va14 NKT cells. *J. Immunol*, 2005, 175, 3648-3655.

[25]   Christian, M; Ralph, M, S; and Shin-ichiro, F; Dendritic cell maturation by innate lymphocytes: coordinated stimulation of innate and adaptive immunity. *J.E.M*, 2005, 202, 203–207.

[26]   Satoshi, K; Ken-ichiro, S; Michishige H; Hiroshi W; Hiroshi, W; Induction of Regulatory Properties in Dendritic Cells by Vα14NKT Cells1. *J. Immunol*, 2005, 175, 3648–3655.

[27]   Thierry, M; Jean, P, Z; Christelle, F; Josette, Fontaine; Emmanuel, Maes; , Frances, Platt; Monique, C; Maria, Leite-de-Moraes; and François, T. Activation of Invariant NKT Cells by the Helminth Parasite Schistosoma mansoni, *J. Immunol*, 2006, 176, 2476-2485.

[28]   Cheng, L. *et al*. Efficient activation of Vα14 invariant NKT cells by foreign lipid antigen is associated with concurrent dendritic cell-specific self recognition. *J. Immunol*. 2007, 178, 2755–2762.

[29]   Haraguchi, K; Takahashi, T; Hiruma, K; Kanda, Y; Tanaka, Y; Ogawa, S; Chiba, S; Miura, O; Sakamaki, H; Hirai, H. Recovery of Valpha24+ NKT cells after hematopoietic stem cell transplantation. *Bone Marrow Transplant.*, 2004, 34, 595-602.

[30]   Seino, K; Taniguchi, M. Functionally distinct NKT cell subsets and subtypes. *J. Exp. Med*. 2005, 202, 1623-1626.

[31]   Taniguchi, M; Harada, M; Kojo, S; Nakayama, T; Wakao, H. The regulatory role of Va14 NKT cells in innate and acquired immune response. *Annu. Rev. Immunol*, 2003, 21, 483-513.

[32]   Kronenberg, M. Toward an understanding of NKT cell biology: progress and paradoxes. *Annu. Rev. Immunol*, 2005, 26, 877-900.

[33]   Yu, K, O, A; Porcelli, S, A. The diverse functions of CD1d-restricted NKT cells and their potential for immunotherapy. *Immunology Letters*, 2005, 100, 42–55.

[34]   Henry, L; Mie, N; Vladislav, R and Andrew J, Nicola. Analysis of the effect of different NKT cell subpopulations on the activation of CD4 and CD8 T cells, NK cells, and B cells. Experimental Hematology, 2006, 34, 289-295.

[35]   David, P; Mar, V; Nasim, M; Sharon, C, W; Susan, E, C and Huge, R. CD161 (Human NKR-P1A) Signaling in NK Cells Involves the Activation of Acid Sphingomyelinase. *J. Immunol*, 2006, 176, 2397–2406.

[36]   Godfrey, D, I & Kronenberg, M. Going both ways: immune regulation via CD1d-dependent NKT cells. *J. Clin. Invest.* 2004, 114,1379–1388.

[37]   Everett, H; Meyer, S, G; Omid, Akbari, Gerald, J, Berry; Paul, B, Savage; Mitchell, K; Toshinori, N; Rosemarie H. D; and Dale, T, U. Glycolipid activation of invariant T

cell receptor[+] NKT cells is sufficient to induce airway hyperreactivity independent of conventional CD4[+] T cells. *PNAS,* 2006, 103, 2782–2787.

[38]   Umetsu, DT; DeKruyff, RH. A role for natural killer T cells in asthma. *Nat. Rev. Immunol,* 2006, 6, 953-958.

[39]   Diego, PG; Paola, PB; Giorgio, PL; Alessandro, B; Michela, R; Margherita, M; Francesco, S; Attilio, BL. CC chemokine receptor expression in childhood asthma is influenced by natural allergen exposure. *Pediatr, Allergy, Immunol,* 2006, 17, 495-500.

[40]   Adcock, IM; Caramori, G. Chemokine receptor inhibitors as a novel option in treatment of asthma. *Curr. Drug. Targets Inflamm.Allergy*, 2004, 3, 257-61.

[41]   Kim, YK; Oh, HB; Lee, EY; Gho, YS; Lee, JE; Kim, YY. Association between a genetic variation of CC chemokine receptor-2 and atopic asthma. *Allergy,* 2007, 62, 208-9.

[42]   Sandford, AJ; Zhu, S; Bai, TR; Fitzgerald, JM; Paré, PD. The role of the C-C chemokine receptor-5 Delta32 polymorphism in asthma and in the production of regulated on activation, normal T cells expressed and secreted. *J. Allergy. Clin. Immunol,* 2001, 108, 69-73.

[43]   Everett H, M; Sho, Goya; Omid, A; Gerald, J, B; Paul, B, S; Mitchell, K; Toshinori, N; Rosemarie, H, D; and Dale, T, U. Glycolipid activation of invariant T cell receptor[+] NKT cells is sufficient to induce airway hyperreactivity independent of conventional CD4[+] T cells. *PNAS,* 2006, 103, 2782–2787.

[44]   Garcia, G; Humbert, M; Capel, F; Rimaniol, AC; Escourrou, P; Emilie, D; Godot, V. Chemokine receptor expression on allergen-specific T cells in asthma and allergic bronchopulmonary aspergillosis. *Allergy,* 2007, 62, 170-7.

[45]   van, Dieren, JM; van, der, Woude, CJ; Kuipers, EJ; Escher, JC; Samsom, JN; Blumberg, RS, Nieuwenhuis; EE. Roles of CD1d-restricted NKT cells in the intestine. *Inflamm. Bowel. Dis*, 2007, 13, 1146-52.

[46]   Fuss, IJ; Heller, F; Boirivant, M; Leon, F; Yoshida, M; Fichtner, S; Yang, Z; Exley, M; Kitani, A; Blumberg, RS; Mannon, P; Strober, W. Nonclassical CD1d-restricted NK T cells that produce IL-13 characterize an atypical Th2 response in ulcerative colitis. *J. Clin. Invest.,* 2004, 113, 1490-7.

[47]   Snyder-Cappione, JE; Nixon, DF; Loo, CP; Chapman, JM; Meiklejohn, DA; Melo, FF; Costa, PR; Sandberg, JK; Rodrigues, DS; Kallas, EG. Individuals with pulmonary tuberculosis have lower levels of circulating CD1d-restricted NKT cells. *.J. Infect. Dis.* 2007, 195, 1361-4.

[48]   Dieli, F; Taniguchi, M; Kronenberg, M; Sidobre, S; Ivanyi, J; Fattorini, L; Iona, E; Orefici, G; De, Leo, G; Russo, D; Caccamo, N; Sireci G; Di, Sano, C; Salerno, A. An anti-inflammatory role for V alpha 14 NK T cells in Mycobacterium bovis bacillus Calmette-Guérin-infected mice. *J Immunol,* 2003, 171, 1961-8.

[49]   Siobhán, C; Cowley; Elizabeth, Hamilton; Jeffrey, A; Frelinger; Jie, Su; James, Forman; Karen, L, Elkins. *CD4-CD8-T cells control intracellular bacterial infections both in vitro and in vivo. JEM,* 2005, 202, 309–319.

[50]   Melanie, J; Scotta, b, c, J; Jason, H; Matthias, T; Dustin, R;. Woodsb and William, G; Cheadle.Interleukin-10 suppresses natural killer cell but not natural killer T cell activation during bacterial infection Cytokine , 22006, 33,79-86.

[51]   Barcelos, W; Martins-Filho, OA; Guimarães, TM. Oliveira, MH; Spíndola-de-Miranda, S; Carvalho, BN. Toledo Vde P Peripheral blood mononuclear cells immunophenotyping in pulmonary tuberculosis patients before and after treatment. *Microbiol Immunol*, 2006, 50, 597-605.

[52]   Tsutsui, H; Adachi, K; Seki, E; Nakanishi, K. Cytokine-induced inflammatory liver injuries. *Curr. Mol. Med*, 2003, 3, 545-59.

[53]   Dong, Z; Zhang, J; Sun, R; Wei, H; Tian, Z. Impairment of liver regeneration correlates with activated hepatic NKT cells in HBV transgenic mice. *Hepatology*, 2007, 45, 1400-12.

[54]   Ahmad, A; Alvarez, F. Role of NK and NKT cells in the immunopathogenesis of HCV-induced hepatitis. *J Leukoc Biol*, 2004, 76, 743-59.

[55]   Biburger, M; Tiegs, G. Alpha-galactosylceramide-induced liver injury in mice is mediated by TNF-alpha but independent of Kupffer cells. *J Immunol*, 2005, 175, 1540-50.

[56]   Saito, T; Okumura, A; Watanabe, H; Asano, M; Ishida-Okawara, A; Sakagami, J; Sudo, K; Hatano-Yokoe, Y; Bezbradica, JS; Joyce, S; Abo, T; Iwakura; Y; Suzuki, K; Yamagoe, S. Increase in hepatic NKT cells in leukocyte cell-derived chemotaxin 2-deficient mice contributes to severe concanavalin A-induced hepatitis. *J. Immunol*, 2004, 173, 579-85.

[57]   McKallip, RJ; Fisher, M; Gunthert, U; Szakal, AK; Nagarkatti, PS; Nagarkatti, M. Role of CD44 and its v7 isoform in staphylococcal enterotoxin B-induced toxic shock: CD44 deficiency on hepatic mononuclear cells leads to reduced activation-induced apoptosis that results in increased liver damage. *Infect Immun*, 2005, 73, 50-61.

[58]   Geissmann, F; Cameron, TO; Sidobre, S; Manlongat, N; Kronenberg, M; Briskin, MJ; Dustin, ML; Littman, DR. Intravascular immune surveillance by CXCR6+ NKT cells patrolling liver sinusoids. *PLoS Biol*, 2005 Apr;3(4):e113. Epub 2005 Apr 5.

[59]   Thorén, F; Romero, A; Lindh, M; Dahlgren, C; Hellstrand, K. A hepatitis C virus-encoded, nonstructural protein (NS3) triggers dysfunction and apoptosis in lymphocytes: role of NADPH oxidase-derived oxygen radicals. *J Leukoc Biol*, 2004, 76, 1180-6.

[60]   Sun, R; Tian, Z; Kulkarni, S. Gao B IL-6 prevents T cell-mediated hepatitis via inhibition of NKT cells in CD4+ T cell- and STAT3-dependent manners. *J. Immunol*, 2004 172, 5648-55.

[61]   de Lalla, C; Galli, G; Aldrighetti, L; Romeo, R; Mariani, M; Monno, A; Nuti, S; Colombo, M; Callea, F; Porcelli, SA; Panina-Bordignon, P; Abrignani, S; Casorati, G; Dellabona, P. Production of profibrotic cytokines by invariant NKT cells characterizes cirrhosis progression in chronic viral hepatitis. *J. Immunol*, 2004, 173, 1417-25.

[62]   Kawamura, H; Aswad, F; Minagawa, M; Govindarajan, S; Dennert, G. P2X7 receptors regulate NKT cells in autoimmune hepatitis. *J. Immunol*, 2006, 176, 2152-60.

[63]   Takeda, K; Hayakawa, Y; Van Kaer, L; Matsuda, H; Yagita, H; Okumura, K. Critical contribution of liver natural killer T cells to a murine model of hepatitis. *Proc. Natl. Acad. Sci. U.S.A*, 2000, 97, 5498-503.

[64]   Hammond, KJ; Godfrey, DI. NKT cells: potential targets for autoimmune disease therapy? *Tissue Antigens*, 2002, 59, 353-63.

[65]   Falcone, M; Facciotti, F; Ghidoli, N; Monti, P; Olivieri, S; Zaccagnino, L; Bonifacio, E; Casorati, G; Sanvito, F; Sarvetnick, N. Up-regulation of CD1d expression restores the immunoregulatory function of NKT cells and prevents autoimmune diabetes in nonobese diabetic mice. *J. Immunol*, 2004, 172, 5908-16.

[66]   Forestier, C; Molano, A; Im, JS; Dutronc, Y; Diamond, B; Davidson, A; Illarionov, PA; Besra, GS; Porcelli, SA. Expansion and hyperactivity of CD1d-restricted NKT cells during the progression of systemic lupus erythematosus in (New Zealand Black x New Zealand White)F1 mice. *J. Immunol,* 2005, 175, 763-70.

[67]   Kojo, S; Adachi, Y; Keino, H; Taniguchi, M; Sumida, T. Dysfunction of T cell receptor AV24AJ18+, BV11+ double-negative regulatory natural killer T cells in autoimmune diseases. *Arthritis Rheum*, 2001, 44, 1127-38.

[68]   Mitsuo, A; Morimoto, S; Nakiri, Y; Suzuki, J; Kaneko, H; Tokano, Y; Tsuda, H; Takasaki, Y; Hashimoto, H. Decreased CD161+CD8+ T cells in the peripheral blood of patients suffering from rheumatic diseases. *Rheumatology*, 2006, 45, 1477-84.

[69]   Fainaru, O; Shseyov, D; Hantisteanu, S; Groner, Y. Accelerated chemokine receptor 7-mediated dendritic cell migration in Runx3 knockout mice and the spontaneous development of asthma-like disease. *Proc. Natl. Acad. Sci. U.S.A*, 2005, 102, 10598-603.

[70]   Lalazar, G; Preston, S; Zigmond, E; Ben Yáacov, A; Ilan, Y. Glycolipids as immune modulatory tools. *Mini. Rev. Med. Chem*, 2006, 6, 1249-53.

[71]   Major, AS; Singh, RR; Joyce, S; Van Kaer, L. The role of invariant natural killer T cells in lupus and atherogenesis. *Immunol Res*, 2006, 34, 49-66.

[72]   Komori, H; Furukawa, H; Mori, S; Ito, MR; Terada, M; Zhang, MC; Ishii, N; Sakuma, N; Nose, M; Ono, M. A signal adaptor SLAM-associated protein regulates spontaneous autoimmunity and Fas-dependent lymphoproliferation in MRL-Faslpr lupus mice. *J. Immunol*, 2006, 176, 395-400.

[73]   Novak, J; Griseri, T; Beaudoin, L; Lehuen, A. Regulation of type 1 diabetes by NKT cells. *Int. Rev. Immunol*, 2007, 26, 49-72.

[74]   Fletcher, JM; Jordan, MA; Baxter, AG. Type 1 Diabetes and NKT Cells: A Report on the 3rd International Workshop on NKT Cells and CD1-Mediated Antigen Presentation. *Rev. Diabet. Stud*, 2004, 1, 141-4.

[75]   Fletcher, JM; Jordan, MA; Baxter, AG. Type 1 Diabetes and NKT Cells: A Report on the 3rd International Workshop on NKT Cells and CD1-Mediated Antigen Presentation. *Rev. Diabet. Stud*, 2004, 1, 141-4.

[76]   Berzins, SP; Cochrane, AD; Pellicci, DG; Smyth, MJ; Godfrey, DI. Limited correlation between human thymus and blood NKT cell content revealed by an ontogeny study of paired tissue samples. *Eur J. Immunol*, 2005, 35, 1399-407.

[77]   Oikawa, Y; Shimada, A; Yamada, S; Motohashi, Y; Nakagawa, Y; Irie, J; Maruyama, T; Saruta, T. NKT cell frequency in Japanese type 1 diabetes. *Ann. N. Y. Acad Sci,* 2003, 1005, 230-2.

[78]   Oikawa, Y; Shimada, A; Yamada, S; Motohashi, Y; Nakagawa, Y; Irie, J; Maruyama, T; Saruta, T. High frequency of valpha24(+) vbeta11(+) T-cells observed in type 1 diabetes. *Diabetes Care,* 2002, 25, 1818-23.

[79] Novak, J; Beaudoin, L; Griseri, T; Lehuen, A. Inhibition of T cell differentiation into effectors by NKT cells requires cell contacts. *J. Immunol,* 2005, 174, 1954-61.

[80] Cardell, SL.The natural killer T lymphocyte: a player in the complex regulation of autoimmune diabetes in non-obese diabetic mice. *Clinical and Experimental Immunology,* 2005, 143, 194–202

[81] Kent, SC; Chen, Y; Clemmings, SM; Viglietta, V; Kenyon, NS; Ricordi, C; Hering, B; Hafler, DA. Loss of IL-4 secretion from human type 1a diabetic pancreatic draining lymph node NKT cells. *J. Immunol,* 2005, 175, 4458-64.

[82] Ly, D; Mi, QS; Hussain, S; Delovitch, TL. Protection from type 1 diabetes by invariant NK T cells requires the activity of CD4+CD25+ regulatory T cells. *J. Immunol,* 2006, 177, 3695-3704.

[83] Wu, L; Van Kaer, L. NKT cells and diabetes. *Am J Physiol Gastrointest Liver Physiol,* 2007. [Epub. ahead of print]

[84] Griseri, T; Beaudoin, L; Novak, J; Mars, LT; Lepault, F; Liblau, R; Lehuen, A. Invariant NKT cells exacerbate type 1 diabetes induced by CD8 T cells. *J. Immunol,* 2005, 175, 2091-101.

[85] Kis, J; Engelmann, P; Farkas, K; Richman, G; Eck, S; Lolley, J; Jalahej, H; Borowiec, M; Kent, SC; Treszl, A; Orban, T. Reduced CD4+ subset and Th1 bias of the human iNKT cells in Type 1 diabetes mellitus. *J. Leukoc Biol,* 2007, 81, 654-62.

[86] Araujo, LM; Lefort, J; Nahori, MA; Diem, S; Zhu, R; Dy, M; Leite-de-Moraes, MC; Bach, JF; Vargaftig, BB; Herbelin, A. Exacerbated Th2-mediated airway inflammation and hyperresponsiveness in autoimmune diabetes-prone NOD mice: a critical role for CD1d-dependent NKT cells. *Eur. J. Immunol,* 2004, 34, 327-35.

[87] Araujo, LM; Lefort, J; Nahori, MA; Diem, S; Zhu, R; Dy, M; Leite-de-Moraes, MC; Bach, JF; Vargaftig, BB; Herbelin, A. Exacerbated Th2-mediated airway inflammation and hyperresponsiveness in autoimmune diabetes-prone NOD mice: a critical role for CD1d-dependent NKT cells. *Eur. J. Immunol,* 2004, 34, 327-35.

[88] Pillai, AB; George, TI; Dutt, S; Teo, P; Strober, S. Host NKT cells can prevent graft-versus-host disease and permit graft antitumor activity after bone marrow transplantation. *J. Immunol,* 2007, 178, 6242-51.

[89] Xiaofeng, Jiang; Takeshi, Shimaoka;Satoshi, Kojo; Michishige, Harada; Hiroshi, Watarai; Hiroshi, Wakao;Nobuhiro, Ohkohchi; Shin, Yonehara; Masaru, Taniguchi; Ken-ichiro Seino; Cutting Edge. Critical Role of CXCL16/CXCR6 in NKT Cell Trafficking in Allograft Tolerance. *The Journal of Immunology,* 2005, 175, 2051–2055.

[90] Galante, NZ; Ozaki, KS; Cenedeze, MA; Kallás, EG; Salomão, R; Pacheco-Silva, A; Câmara, NO. Frequency of Valpha24+Vbeta11+ NKT cells in peripheral blood of human kidney transplantation recipients. *Int. Immunopharmacol,* 2005, 5, 53-8.

[91] Li, Y; Koshiba, T; Yoshizawa, A; Yonekawa, Y; Masuda, K; Ito, A; Ueda, M; Mori, T; Kawamoto, H; Tanaka, Y; Sakaguchi, S; Minato, N; Wood, KJ; Tanaka, K. Analyses of peripheral blood mononuclear cells in operational tolerance after pediatric living donor liver transplantation. *Am. J. Transplant,* 2004. 4, 2118-25.

[92] Westerhuis, G; Maas, WG; Willemze, R; Toes, RE; Fibbe, WE. Long-term mixed chimerism after immunologic conditioning and MHC-mismatched stem-cell transplantation is dependent on NK-cell tolerance. *Blood,* 2005, 106, 2215-20.

[93]   Haraguchi, K; Takahashi, T; Hiruma, K; Kanda, Y; Tanaka, Y; Ogawa, S; Chiba, S; Miura, O; Sakamaki, H; Hirai, H. Recovery of Valpha24+ NKT cells after hematopoietic stem cell transplantation. *Bone Marrow Transplant*, 2004, 34, 595-602.

[94]   Kawamura, H; Kameyama, H; Kosaka, T; Kuwahara, O; Bannai, M; Kawamura, T; Watanabe, H; Abo, T. Association of CD8+ natural killer T cells in the liver with neonatal tolerance phenomenon. *Transplantation*, 2002, 73, 978-92.

[95]   Masaki, Terabe;Jeremy, Swann; Elena, Ambrosino; Pratima, Sinha; Shun, Takaku; Yoshihiro, Hayakawa; Dale, I; Godfrey; Suzanne, Ostrand-Rosenberg; Mark, J; Smyth; Jay, A. Berzofsky,A nonclassical non-Vα14Jα18 CD1d-restricted (type II) NKT cell is sufficient for downregulation of tumor immunosurveillance. *J. E. M*, 2005, 202, 1627–1633

[96]   Karl, O, A, Yu; Steven, A; Porcelli. The diverse functions of CD1d-restricted NKT cells and their potential for immunotherapy. *Immunology Letters*, 2005, 100, 42–55.

[97]   van, Dieren, JM; van, der, Woude, CJ; Kuipers, EJ; Escher, JC; Samsom, JN; Blumberg, RS. Roles of CD1d-restricted NKT cells in the intestine. *Nieuwenhuis EE. Inflamm. Bowel. Dis,* 2007,13,1146-52.

[98]   Sen, Y; Yongyi, B; Yuling, H; Luokun, X; Li, H; Jie, X; Tao, D; Gang, Z; Junyan, L; Chunsong, H; Zhang, X; Youxin, J; Feili, G; Boquan, J; Jinquan, T. V alpha 24-invariant NKT cells from patients with allergic asthma express CCR9 at high frequency and induce Th2 bias of CD3+ T cells upon CD226 engagement. *J. Immunol*, 175, 4914–4926.

[99]   Akbari, O; Faul, J, L; Hoyte, E, G; Berry, G, J; Wahlstrom, J; Kronenberg, M; DeKruyff, R, H; Umetsu, D, T. CD4+ invariant T-cell-receptor+ natural killer T cells in bronchial asthma. *N. Engl. J. Med*, 2006, 354, 1117–1129.

[100]   Kronenberg, M. Toward an understanding of NKT cell biology: progress and paradoxes. *Annu. Rev. Immunol*, 2005, 23, 877-900.

[101]   Joyce, S; Van Kaer, L. CD1-restricted antigen presentation: an oily matter. *Curr. Opin. Immunol*, 2003, 15, 95-104.

[102]   Kjer-Nielsen, L. A structural basis for selection and cross-species reactivity of the semi-invariant NKT cell receptor in CD1d/glycolipid recognition. *.J. Exp. Med*, 2006, 203, 661-673.

[103]   Schümann, J; Voyle, R, B; Wei, B, Y; & MacDonald, H, R. Cutting edge: influence of the TCR Vβ domain on the avidity of CD1d: α-galactosylceramide binding by invariant Vα14 NKT cells. *J. Immunol.* 2003, 170, 5815-5819.

[104]   Dougan, SK; Salas, A; Rava, P; Agyemang, A; Kaser, A; Morrison, J; Khurana, A; Kronenberg, M; Johnson, C. Microsomal triglyceride transfer protein lipidation and control of CD1d on antigen-presenting cells. *J. Exp. Med*, 2005, 202, 529-539.

[105]   Joyce, S; Natural ligand of mouse CD1d1: cellular glycosylphosphatidylinositol. *Science*, 1998, 279, 1541-1544.

[106]   Dougan, SK; Rava, P; Hussain, M & Blumberg, RS. MTP regulated by an alternate promoter is essential for NKT cell development. *J. Exp. Me*, 2007, 204, 533-545.

[107]   Sagiv, Y; A distal effect of microsomal triglyceride transfer protein deficiency on the lysosomal recycling of CD1d. *J. Exp. Med*, 2007, 204, 921-928.

[108]   Major, AS; Joyce, S; Van, Kaer, L. Lipid metabolism, atherogenesis and CD1-restricted antigen presentation. *Trends Mol. Med*, 2006, 12, 270-278.

[109] De Libero, G; Mori, L. Mechanisms of lipid-antigen generation and presentation to T cells. *Trends Immunol,* 2006, 27, 485-492.

[110] van, den, Elzen, P; Garg, S; Leon, L; Brigl, M; Leadbetter, EA; Gumperz, JE; Dascher, CC; Cheng, TY; Sacks, FM; Illarionov, PA. Apolipoprotein-mediated pathways of lipid antigen presentation. *Nature,* 2005, 437, 906-910.

[111] Luc Van Kaer. NKT cells: T lymphocytes with innate effector functions. *Curr. Opin. Immunol,* 2007, 19, 354-364.

[112] Schrantz, N. The Niemann-Pick type C2 protein loads isoglobotrihexosylceramide onto CD1d molecules and contributes to the thymic selection of NKT cells. *J. Exp. Med,* 2007, 204, 841-852.

[113] Zhou, D. Editing of CD1d-bound lipid antigens by endosomal lipid transfer proteins. *Science,* 2004, 303, 523-527

[114] Exley, M; Garcia, J; Balk, S, P & Porcelli, S. Requirements for CD1d recognition by human invariant Vα24+ CD4–CD8– T cells. *J. Exp. Med.,* 1997, 186, 109-120.

[115] Brigl, M; Bry, L.; Kent, S, C; Gumperz, J, E & Brenner, M, B. Mechanism of CD1d-restricted natural killer T cell activation during microbial infection. *Nature, Immunol,* 2003, 4, 1230-1237.

[116] Mattner, J. et al. Exogenous and endogenous glycolipid antigens activate NKT cells during microbial infections. *Nature,* 2005, 434, 525-529.

[117] Zhou, D. et al. Lysosomal glycosphingolipid recognition by NKT cells. *Science,* 2004, 306, 1786-1789.

[118] Porubsky, S; Speak, AO; Luckow, B; Cerundolo, V; Platt, FM; Gro, ne, H-J. Normal development and function of invariant natural killer T cells in mice with isoglobotrihexosylceramide(iGb3) deficiency. *Proc. Natl. Acad. Sci. USA,* 2007, 104, 5977-5982.

[119] Wei, D, G; Curran, S, A; Savage, P, B; Teyton, L & Bendelac, A. Mechanisms imposing the Vβ bias of Vα14 natural killer T cells and consequences for microbial glycolipid recognition. *J. Exp. Med.,* 2006, 203, 1197-1207.

[120] De Libero, G. et al. Bacterial infections promote T cell recognition of self-glycolipids. *Immunity,* 2005, 22, 763–772.

[121] Berntman, E; Rolf, J; Johansson, C; Anderson, P &Cardell, S, L. The role of CD1d-restricted NKT lymphocytes in the immune response to oral infection with Salmonella typhimurium. *Eur. J. Immunol,* 2005, 35, 2100-2109.

[122] Skold, M.; Xiong, X; Illarionov, P, A; Besra, G, S & Behar, S, M. Interplay of cytokines and microbial signals in regulation of CD1d expression and NKT cell activation. *J. Immunol..,* 2005, 175, 3584-3593.

[123] Raghuraman, G; Geng, Y & Wang, C, R. IFN--mediated up-regulation of CD1d in bacteria-infected APCs. *J. Immunol,* 2006, 177, 7841-7848.

[124] Kinjo, Y; Wu, DY; Kim, G; Xing, GW; Poles, MA; Ho, DD; Tsuji, M; Kawahara, K; Wong, CH; Kronenberg, M. Recognition of bacterial glycosphingolipids by natural killer T cells. *Nature,* 2005, 434, 520-525.

[125] Faveeuw, C. et al. Antigen presentation by CD1d contributes to the amplification of Th2 responses to Schistosoma mansoni glycoconjugates in mice. *J. Immunol,* 2002, 169, 906-912.

[126]   Schumann, J; Mycko, M., P; Dellabona, P; Casorati, G & Macdonald, H, R. Cutting edge: influence of the TCR Vβ domain on the selection of semi-invariant NKT cells by endogenous ligands. *J. Immunol,* 2006, 176, 2064-2068.

[127]   Mallevaey, T. et al. Activation of invariant NKT cells by the helminth parasite schistosoma mansoni. *J. Immunol,* 2006, 176, 2476-2485.

[128]   Zhou, D. et al. Lysosomal glycosphingolipid recognition by NKT cells. *Science,* 2004, 306, 1786-1789.

[129]   Gadola, S, D. et al. Impaired selection of invariant natural killer T cells in diverse mouse models of glycosphingolipid lysosomal storage diseases. *J. Exp., Med,* 2006, 203, 2293-2303.

[130]   Schümann, J. et al. Differential alteration of lipid antigen presentation to NKT cells due to imbalances in lipid metabolism. *Eur. J. Immunol,* 2007, 37, 1431–1441.

[131]   Porubsky, S. *et al.* Normal development and function of invariant natural killer T cells in mice with isoglobot rihexosylceramide (iGb3) deficiency. *Pro. Natl. Acad. Sci, U.S.A,* 2007, 104, 5977-5982.

[132]   Speak, A, O. *et al.* Implications for invariant natural killer T cell ligands due to the restricted presence of isoglobotrihexosylceramide in mammals. *Proc. Natl. Acad. Sci. U.S.A,* 2007, 104, 5971-5976.

[133]   Xia, C. *et al.* Synthesis and biological evaluation of α-galactosylceramide (KRN7000) and isoglobotrihexosylceramide (iGb3). *Bioorg. Med. Chem. Lett,* 2006, 16, 2195-2199.

[134]   Xia, C. *et al.* Thio-isoglobotrihexosylceramide, an agonist for activating invariant natural killer T cells. *Org. Lett,* 2006, 8, 5493-5496.

[135]   Kronenberg, M & Gapin, L. Natural killer T cells: know thyself. *Proc. Natl. Acad. Sci. U. S. A,* 2007, 104, 5713-5714.

[136]   Mattner, J; DeBord, KL; Ismail, N; Goff, RD; Cantu, CI; Zhou, D; Saint, Mezard, P; Wang, V; Gao, Y; Yin, N. Exogenous and endogenous glycolipid antigens activate NKT cells during microbial infections. *Nature,* 2005, 434, 525-529.

[137]   Nagarajan, N, A & Kronenberg, M. Invariant NKT cells amplify the innate immune response to lipopolysaccharide. *J. Immunol.,* 2007, 178, 2706-2713.

[138]   Montoya, C, J. et al. Activation of plasmacytoid dendritic cells with TLR9 agonists initiates invariant NKT cell-mediated cross-talk with myeloid dendritic cells. *J. Immunol,* 2006, 177, 1028-1039.

[139]   Schofield, L. et al. CD1d-restricted immunoglobulin G formation to GPI-anchored antigens mediated by NKT cells. *Science,* 1999, 283, 225-229.

[140]   Molano, A. et al. Cutting edge: the IgG response to the circumsporozoite protein is MHC class II-dependent and CD1d-independent: exploring the role of GPIs in NK T cell activation and antimalarial responses. *J. Immunol.,* 2000, 164, 5005-5009.

[141]   Romero, J, F; Eberl, G; MacDonald, H, R. & Corradin, G. CD1d-restricted NK T cells are dispensable for specific antibody responses and protective immunity against liver stage malaria infection in mice. *Parasite. Immunol,* 2001, 23, 267-269.

[142]   Amprey, J, L. et al. A subset of liver NK T cells is activated during Leishmania donovani infection by CD1d-bound lipophosphoglycan. *J. Exp. Med,* 2004, 200, 895-904.

[143]   Fischer, K. et al. Mycobacterial phosphatidylinositol mannoside is a natural antigen for CD1d-restricted T cells. *Proc. Natl. Acad. Sci. U.S.A,* 2004, 101, 10685-10690.

[144] Kinjo, Y. et al. Natural killer T cells recognize diacylglycerol antigens from pathogenic bacteria. *Nature Immunol*, 2006, 7, 978-986.

[145] Kinjo, Y. et al. Recognition of bacterial glycosphingolipids by natural killer T cells. *Nature*, 2005, 434, 520-525.

[146] Sriram, V; Du, W; Gervay, Hague, J. & Brutkiewicz, R, R. Cell wall glycosphingolipids of Sphingomonas paucimobilis are CD1d-specific ligands for NKT cells. *Eur. J. Immunol.*, 2005, 35, 1692–701.

[147] Kawahara, K; Moll, H; Knirel, Y, A; Seydel, U. & Zahringer, U. Structural analysis of two glycosphingolipids from the lipopolysaccharide-lacking bacterium Sphingomonas capsulata. *Eur. J. Biochem*, 2000, 267, 1837-1846.

[148] Kawahara, K.; Kubota, M.; Sato, N; Tsuge, K. & Seto, Y. Occurrence of an □-galacturonosyl-ceramide in the dioxin-degrading bacterium Sphingomonas wittichii. *FEMS Microbiol Lett*, 2002, 214, 289-294.

[149] Perola, O. et al. Recurrent Sphingomonas paucimobilis-bacteraemia associated with a multi-bacterial waterborne epidemic among neutropenic patients. *J. Hosp. Infect*, 2002, 50, 196-201.

[150] Orloski, K, A, Hayes, E, B; Campbell, G, L & Dennis, D, T. Surveillance for Lyme disease. *MMWR.. CDC. Surveill*, 2000, 49, 1-11.

[151] Hammond, KJL; Poulton, LD; Palmisano, LJ; Silveira, PA; Godfrey, DI and Baxter, AG. -T cell receptor (TCR) [+]CD4[-]CD8[-] (NKT) thymocytes prevent insulin-dependent diabetes mellitus in nonobese diabetic (NOD)/Lt mice by the influence of interleukin (IL)-4 and/or IL-10. *J. Exp. Med*, 1998, 187, 1047-1056.

[152] Laloux, V; Beaudoin, L; Jeske, D; Carnaud, C and Lehuen, A. NK T cell-induced protection against diabetes in V alpha 14-J alpha 281 transgenic nonobese diabetic mice is associated with a Th2 shift circumscribed regionally to the islets and functionally to islet autoantigen. *J. Immunol.*, 2001, 166, 3749-3756.

[153] Van Kaer, L. alpha-Galactosylceramide therapy for autoimmune diseases: prospects and obstacles. *Nat. Rev. Immunol.*, 2005, 5, 31-42.

[154] Forestier, C; Takaki, SA; Molano, A; Im, JS; Baine, I; Jerud, ES; Illarionov, P; Ndonye, R; Howell, AR; Santamaria, P; Besra, GS; DiLorenzo, TP and Porcelli, S. Improved outcomes in NOD mice treated with a novel Th2 cytokine-biasing NKT cell activator. *J. Immunol.*, 200, 178, 1415-1425.

[155] Mizuno, M; Masumura, M; Tomi, C; Chiba, A; Oki, S; Yamamura, T and Miyake, S. Synthetic glycolipid OCH prevents insulitis and diabetes in NOD mice. *J. Autoimmun*, 2004, 23, 293-300.

[156] Mi, Q, S; Ly, D; Zucker, P; McGarry, M and Delovitch, TL. Interleukin-4 but not interleukin-10 protects against spontaneous and recurrent type 1 diabetes by activated CD1d-restricted invariant natural killer T cells. *Diabetes*, 2004, 53, 1303-1310.

[157] Chen, YG; Chen, J; Osborne, MA; Chapman, HD; Besra, GS; Porcelli, SA; Leiter, EH; Wilson, SB and Serreze, DV. CD38 is required for the peripheral survival of immunotolerogenic CD4+ invariant NK T cells in nonobese diabetic mice. *J. Immunol.*, 2006, 177, 2939-2947.

[158] Cain, JA; Smith, JA; Ondr, JK; Wang, B and Katz, JD. NKT cells and IFN-gamma establish the regulatory environment for the control of diabetogenic T cells in the nonobese diabetic mouse. *J. Immunol.*, 2006, 176, 1645-1654.

[159]   Chatenoud, L. Do NKT cells control autoimmunity? *J. Clin. Invest,* 2002, 110,793-748.

[160]   Chen, YG; Choisy-Rossi, CM; Holl, TM; Chapman, HD; Besra, GS; Porcelli, SA; Shaffer, DJ; Roopenian, D; Wilson, SB and Serreze, DV. Activated NKT cells inhibit autoimmune diabetes through tolerogenic recruitment of dendritic cells to pancreatic lymph nodes. *J. Immunol.,* 2005, 174, 1196-1204.

[161]   Naumov, YN; Bahjat, KS; Gausling, R; Abraham, R; Exley, MA; Koezuka, Y; Balk, SB; Strominger, JL; Clare-Salzer, M and Wilson, SB. Activation of CD1drestricted T cells protects NOD mice from developing diabetes by regulating dendritic cell subsets. *Proc. Natl. Acad. Sci. U.S.A.,* 2001, 98, 13838-13843.

[162]   Ly, D; Mi, Q□S; Hussain, S and Delovitch, TL. Protection from type 1 diabetes by invariant NK T cells requires the activity of CD4+CD25+ regulatory T cells. *J. Immunol.,* 2006, 177, 3695-3704.

[163]   Beaudoin, L; Laloux, V; Novak, J; Lucas, B and Lehuen, A. NKT cells inhibit the onset of diabetes by impairing the development of pathogenic T cells specific for pancreatic cells. *Immunity,* 2002, 17, 725-736.

[164]   MacDonald, HR; Mycko, MP. Development and selection of Valpha l4i NKT cells. *Curr. Top. Microbiol. Immunol.,* 2007, 314, 195-212.

[165]   Schmieg, J; Yang, G; Franck, RW; Van, Rooijen, N and Tsuji, M. Glycolipid presentation to natural killer T cells differs in an organ-dependent fashion. *Proc. Natl. Acad. Sci. U.S.A.,* 2005, 102, 1127-1132.

[166]   Godfrey, DI; MacDonald, HR; Kronenberg, M; Smyth, MJ and Van, Kaer, L; NKT cells: what's in a name? *Nat. Rev. Immunol.,* 2004, 4, 231-237.

[167]   Li, C; Bai, X; Wang, S; Tomiyama-Miyaji, C; Nagura, T; Kawamura, T; Abo, T. Immunopotentiation of NKT cells by low-protein diet and the suppressive effect on tumor metastasis. *Cell, Immunol.,* 2004, 231, 96-102.

[168]   Crowe, NY; Coquet, JM; Berzins, SP; Kyparissoudis, K; Keating, R; Pellicci, DG; Hayakawa, Y; Godfrey, DI; Smyth, MJ. Differential antitumor immunity mediated by NKT cell subsets in vivo. *J. Exp. Med.* 2005, 202, 1279-1288.

[169]   Masaki, Terabe; Jeremy, Swann; Elena, Ambrosino; Pratima, Sinha; Shun, Takaku; Yoshihiro, Hayakawa; Dale, I. Godfrey; Suzanne, Ostrand-Rosenberg; Mark, J. Smyth and Jay, A. Berzofsky. A nonclassical non-Vα14Jα18 CD1d-restricted (type II) NKT cell is sufficient for down regulation of tumor immunosurveillance. *J.E.M.,* 2005, December 19, Vol. 202, No. 12, 1627–1633.

[170]   Kirsten, JL□ Hammond and Mitchell, Kronenberg. Natural killer T cells: natural or unnatural regulators of autoimmunity? *Current Opinion in Immunology,* 2003, 15, 683–689.

[171]   Kojo, S; Adachi, Y; Keino, H; Taniguchi, M; Sumida T Dysfunction of T cell receptor AV24AJ18+, BV11+ double-negative regulatory natural killer T cells in autoimmune diseases. *Arthritis Rheum,* 2001, 44, 1127-38.

[172]   Singh, AK; Yang, JQ; Parekh, VV; Wei, J; Wang, CR; Joyce, S; Singh, RR; Van, Kaer, L. The natural killer T cell ligand alpha-galactosylceramide prevents or promotes pristane-induced lupus in mice. *Eur. J. Immunol.,* 2005, 35, 1143-54.

[173]   eng, Tao; Liu, Shangwu; Wu, Qun; Liu, Yan; Ju, Wei; Liu, Junyan; Gong, Feili; Jin, Boquan and Tan, Jinquan. CD226 Expression Deficiency Causes High Sensitivity to

Apoptosis in NK T Cells from Patients with Systemic Lupus Erythematosus1. *The Journal of Immunology,* 2005, 174, 1281–1290.

[174] Novak, J; Griseri, T; Beaudoin, L; Lehuen, A. Regulation of type 1 diabetes by NKT cells. *Int. Rev. Immunol.* 2007, 26, 49-72.

[175] Cain, JA; Smith, JA; Ondr, J, K; Wang, B; Katz, JD. NKT cells and IFN-gamma establish the regulatory environment for the control of diabetogenic T cells in the nonobese diabetic mouse. *J. Immunol.* 2006, 176, 1645-54.

[176] Jiang, X; Kojo, S; Harada, M; Ohkohchi, N; Taniguchi, M; Seino, KI. Mechanism of NKT cell-mediated transplant tolerance. *Am. J. Transplant.*, 2007, 7, 1482-90.

[177] Yu, KO; Im, JS; Illarionov, PA; Ndonye, RM; Howell, AR; Besra, GS; Porcelli, SA. Production and characterization of monoclonal antibodies against complexes of the NKT cell ligand alpha-galactosylceramide bound to mouse CD1d. *J Immunol Methods Epub,* 2007, 323, 11-23.

[178] Okada, H; Nagamura-Inoue, T; Mori, Y; Takahashi, TA. Expansion of Va24+Vb11+ NKT cells from cord blood mononuclear cells using IL-15, IL-7 and Flt3-L depends on monocytes. *Eur. J. Immunol,* 2006, 36, 236–244.

[179] Lisbonne, M; Diem, S; de, Castro, Keller, A; Lefort, J; Araujo, LM; Hachem, P; Fourneau, JM; Sidobre, S; Kronenberg, M; Taniguchi, M; Van, Endert, P; Dy, M; Askenase, P; Russo, M; Vargaftig, BB; Herbelin, A; Leite-de-Moraes, MC. Cutting edge: invariant V alpha 14 NKT cells are required for allergen-induced airway inflammation and hyperreactivity in an experimental asthma model. *J. Immunol.* 2003, 171, 1637-41.

[180] Dombrowicz, D. Exploiting the innate immune system to control allergic asthma. *Eur. J. Immunol.,* 2005, 35, 2786-8.

[181] Hachem, P; Lisbonne, M; Michel, ML; Diem, S; Roongapinun, S; Lefort, J; Marchal, G; Herbelin, A; Askenase, PW; Dy, M; Leite-de-Moraes, MC. Alpha-galactosylceramide-induced iNKT cells suppress experimental allergic asthma in sensitized mice: role of IFN-gamma. *Eur. J. Immunol.,* 2005, 35, 2793-802.

[182] Ikegami, Y; Yokoyama, A; Haruta, Y; Hiyama, K; Kohno, N. Circulating natural killer T cells in patients with asthma. *J. Asthma.,* 2004, 41, 877-82.

[183] Van, Kaer, L. a-Galactosylceramide therapy for autoimmune diseases: prospects and obstacles. *Nat. Rev. Immunol.,* 2005, 5, 31-42.

[184] Van, Kaer, L. Natural killer T cells as targets for immunotherapy of autoimmune diseases. *Immunol. Cell Biol.,* 2004, 82, 315-22.

[185] Kent, SC; Chen, Y; Clemmings, SM; Viglietta, V; Kenyon, NS; Ricordi, C; Hering, B; Hafler, DA. Loss of IL-4 secretion from human type 1a diabetic pancreatic draining lymph node NKT cells. *J. Immunol,* 2005, 175, 4458-64.

[186] Sharif, S; Arreaza, G, A; Zucker, P; Delovitch, TL. Regulatory natural killer T cells protect against spontaneous and recurrent type 1 diabetes. *Ann. N. Y. Acad. Sci,* 2002, 958, 77-88.

[187] Green, MR; Kennell, AS; Larche, MJ; Seifert, MH; Isenberg, DA; Salaman, MR. Natural killer T cells in families of patients with systemic lupus erythematosus: their possible role in regulation of IGG production. *Arthritis. Rheum,* 2007, 56, 303-10.

[188] Nemoto, K; Mae, T; Sugawara, Y; Hayashi, M; Abe, F; Takeuchi, T. Deoxyspergualin therapy in autoimmune MRL/1pr mice suffering advanced lupus-like disease. *J. Antibiot,* 1990, 43, 1590-6.

[189] Kajiwara, T; Tomita, Y; Okano, S; Iwai, T; Yasunami, Y; Yoshikai, Y; Nomoto, K; Yasui, H; Tominaga, R. Effects of cyclosporin A on the activation of natural killer T cells induced by alpha-galactosylceramide. *Transplantation*, 2007, 83, 184-92.

[190] Li, W; Carper, K; Perkins, J,D. Enhancement of NKT cells and increase in regulatory T cells results in improved allograft survival. *J. Surg. Res*, 2006, 134, 10-21.

[191] Fujii, S. Application of natural killer T-cells to posttransplantation immunotherapy. *Int. J. Hematol*, 2005, 81, 1-5.

[192] Margalit, M; Ilan, Y; Ohana, M; Safadi, R; Alper, R; Sherman, Y; Doviner, V; Rabbani, E; Engelhardt, D; Nagler, A. Adoptive transfer of small numbers of DX5$^+$ cells alleviates graft-versus-host disease in a murine model of semiallogeneic bone marrow transplantation: a potential role for NKT lymphocytes. *Bone Marrow Transplant*, 2005, 35, 191-7.

[193] Sonoda, KH; Taniguchi, M; Stein-Streilein, J. Long-term survival of corneal allografts is dependent on intact CD1d-reactive NKT cells. *J. Immunol.* 2002, 168, 2028-34.

[194] Ikehara, Y; Yasunami, Y; Kodama, S; Maki, T; Nakano, M; Nakayama, T; Taniguchi, M; Ikeda, S. CD4$^+$ Vα14 natural killer T cells are essential for acceptance of rat islet xenografts in mice. *J. Clin. Invest*, 2000, 105, 1761–1767.

[195] Iwai, T; Tomita, Y; Okano, S; Shimizu, I; Yasunami, Y; Kajiwara, T; Yoshikai, Y; Taniguchi, M; Nomoto, K; Yasui, H. Regulatory roles of NKT cells in the induction and maintenance of cyclophosphamide-induced tolerance. *J. Immunol.*, 2006, 177, 8400-9.

[196] Takeda, K; Hayakawa, Y; Van, Kaer, L; Matsuda, H; Yagita, H; Okumura, K. Critical contribution of liver natural killer T cells to a murine model of hepatitis. *Proc. Natl. Acad. Sci. U.S.A*, 2000, 97, 5498-503.

[197] Israeli, E; Goldin, E; Shibolet, O; Klein, A; Hemed, N; Engelhardt, D; Rabbani, E; Ilan, Y; World, J, Gastroenterol. Oral immune regulation using colitis extracted proteins for treatment of Crohn's disease: results of a phase I clinical trial, *World Journal of Gastroenterology*, 2005, 11, 3105-11.

[198] Ueno, Y; Tanaka, S; Sumii, M; Miyake, S; Tazuma, S; Taniguchi, M; Yamamura, T; Chayama, K. Single dose of OCH improves mucosal T helper type 1/T helper type 2 cytokine balance and prevents experimental colitis in the presence of valpha14 natural killer T cells in mice. *Inflamm. Bowel. Dis*, 2005, 11, 35-41.

[199] Krajina, T; Leithäuser, F; Möller, P; Trobonjaca, Z; Reimann, J. Colonic lamina propria dendritic cells in mice with CD4$^+$ T cell-induced colitis. *Eur. J. Immunol*, 2003, 33, 1073-83.

[200] Renukaradhya, GJ; Webb, TJ; Khan, MA; Lin, YL; Du, W; Gervay-Hague, J; Brutkiewicz, RR. Virus-Induced Inhibition of CD1d1-Mediated Antigen Presentation: Reciprocal Regulation by p38 and ERK1. *J. Immunol*, 2005, 175, 4301-4308.

[201] Markus, Moll; Jennifer, Snyder-Cappione; Gerald, Spotts; Frederick, M; Hecht; Johan, K; Sandberg; Douglas, F, Nixon. Expansion of CD1d-restricted NKT cells in patients with primary HIV-1 infection treated with interleukin-2. *Blood*, 2006, 107, 3081-3083

[202] Warren, S, Pearl; LiLi, Tu; Paul, L, Stein. Choosing the T and NKT pathways. *Curr. Opini. Immunol*, 2004, 16, 167–173.

[203] Jason, C; Mercera, b; Melanie, J; Ragina, c; Avery, Augusta, C; Natural killer T cells: rapid responders controlling immunity and disease. *The International Journal of Biochemistry & Cell Biology, 2005,* 37, 1337–1343.

[204] Mitchell, Kronenberg. Progress and Paradoxes. Annu. *Rev. Immunol,* 2005, 26, 877–900.

[205] Kirsten, J,L; Hammond, Mitchell, Kronenberg. natural or unnatural regulators of autoimmunity? *Curr. Opin. Immunol,* 2003, 15, 683–689.

[206] Anderson☐Joseph; Akkina, Ramesh. CXCR4 and CCR5 shRNA transgenic CD34$^+$ cell derived macrophages are functionally normal and resist HIV-1 infection. *Retrovirology*, 2005, 2, 53.

[207] Green, MR; Kennell, AS; Larche, MJ; Seifert, MH; Isenberg, DA; Salaman, MR. Natural killer T cells in families of patients with systemic lupus erythematosus: their possible role in regulation of IGG production. *Arthritis Rheum,* 2007, 56, 303-10.

[208] Luc, Van, Kaer. NKT cells: T lymphocytes with innate effector functions. *Curr. Opin. Immunol,* 2007, 19, 354–364.

Chapter 2

Lifestyle and Natural Killer Activity

Kanehisa Morimoto[1,,#] and Qing Li[2,*]*

[1] Department of Social and Environmental Medicine, Osaka University
Graduate School of Medicine, Suita, Osaka, Japan
[2] Department of Hygiene and Public Health, Nippon Medical School
Tokyo, Japan

Abstract

Objective

Lifestyle factors including cigarette smoking and alcohol consumption are associated with mortality and morbidity due to cancers of various organs. It is well documented that natural killer (NK) cells provide host defense against tumors and viruses. In order to explore whether lifestyle and a forest bathing trip (natural aromatherapy) affects human NK and lymphokine-activated killer (LAK) activities, NK and CD56[+]/CD3[+] NKT cells, intracellular levels of perforin, granzymes, and granulysin, and the underlying mechanism of any effects, we have conducted a series of investigations to clarify this issue.

Methods

Fifty-four to one hundred and fourteen healthy male subjects, aged 20-60 years, from a large company in Osaka, Japan were selected with informed consent. The subjects were divided into groups with good, moderate, and poor lifestyles according to their responses on a questionnaire regarding eight health practices (cigarette smoking, alcohol consumption, sleeping hours, working hours, physical exercise, eating breakfast,

---

[*] Kanehisa Morimoto and Qing Li contributed equally to this work.
[#] E-mail address:morimoto@envi.med.osaka-u.ac.jp, TEL: +81-6-6879-3920, FAX: +81-6-6879-3929

balanced nutrition, and mental stress). Peripheral blood was taken, and NK and LAK activities were measured by ^{51}Cr-release assay, numbers of NK, T, and CD56$^+$/CD3$^+$ NKT cells, and perforin-, granulysin-, and granzymes A/B-expressing cells in peripheral blood lymphocytes (PBL) were measured by flow cytometry. In the forest bathing trips, 12 healthy male subjects aged 35-56 years were selected with informed consent. The subjects experienced three-day/two-night trips to a forest area and to a city without forests, respectively. Blood was sampled on the second and third days during the trips, and on days 7 and 30 after the forest bathing trip. Similar measurements were made before the trips on normal working days as a control. NK activity was measured by ^{51}Cr-release assay, numbers of NK, T, and CD56$^+$/CD3$^+$ NKT cells, and perforin-, granulysin-, and granzymes A/B-expressing cells in PBL were measured by flow cytometry.

Results

Subjects with good or moderate lifestyles showed significantly higher NK and LAK activities, higher numbers of NK cells, and perforin-, granulysin-, and granzymes A/B-expressing cells and a significantly lower number of T cells in PBL than subjects with a poor lifestyle. Among the eight health practices, cigarette smoking, physical exercise, eating breakfast, working hours, and balanced nutrition significantly affected the numbers of NK, T, and CD56$^+$/CD3$^+$ NKT cells, and perforin-, granulysin-, and granzymes A/B-expressing cells, and alcohol consumption significantly affected the number of granzyme A-expressing cells and NKT cells. On the other hand, mental stress and sleeping hours had no effect on these parameters. The forest bathing trip significantly increased NK activity, the number of NK cells, and perforin-, granulysin- and granzymes A/B-expressing cells, and the positive rate of NKT cells. In contrast, a city tourist visit did not show any effects on the above parameters.

Conclusion

Taken together, these findings indicate that good lifestyle practices and forest bathing trips significantly increased human NK and LAK activity, and the numbers of NK and NKT cells, and perforin-, granulysin-, and granzymes A/B-expressing cells in PBL.

Key words: Forest bathing, Granulysin, Granzyme, LAK, Lifestyle, Natural aromatherapy, NK activity, NKT cells, Perforin, smoking

Introduction

Lifestyle factors including cigarette smoking and alcohol consumption are associated with mortality and morbidity due to cancers of various organs (Hirayama, 1990). Lifestyle can be assessed by asking questions on the following eight lifestyle practices and expressed as a health practice index (HPI): amount of smoking and drinking, eating breakfast, sleeping hours, working hours, physical exercise, nutritional balance, and mental stress. These criteria were based on seven health practices demonstrated in an Alameda County (California) study (Berkman and Breslow, 1983) and modified by Morimoto to reflect the Japanese situation more accurately (Morimoto, 1990) (Table 1). It is well known that natural killer (NK) cells improve defense against viruses and some tumors (Imai et al., 2000). During malignancy, NK cells appear to represent a first line of defense against the metastatic spread of blood-borne tumor cells, and NK cell activity may be important in immune surveillance against tumors (Trinchieri 1989). Based on the background mentioned above, we review how lifestyle affects human NK activity and the underlying mechanism in this study.

Table 1. Eight items used to calculate the health practice index (Morimoto, 1990; Kusaka et al., 1992; Inoue et al., 1996; Li et al., 2007a)

	Good health practice (score 1)	Poor health practice (score 0)
Cigarette smoking	Not smoking or quit smoking	Smoking
Alcohol drinking	Sometimes or not	Almost every day
Sleeping hours	7 or 8 hr/night	≤ 6 or $9 \geq$ hr/night
Working hours	≤ 9 hr/day	≥ 10 hr/day.
Physical exercise	Once and more a week	Less than once a week
Eating breakfast	Everyday	Sometimes or not
Nutritional balance	Eating a balanced diet	Do not eat a balanced diet
Mental stress	Feel mild	Feel excessive stress or slight stress

Effect of Lifestyle on Human NK Activity

Among the lifestyle factors, cigarette smoking is the most important factor affecting the immune system. It is well known that smoking is associated with a raised incidence of a variety of malignant diseases such as carcinomas of the lung, esophagus, rectum, pancreas, and bladder (Doll and Peto, 1976). It has also been shown that there is a higher incidence of metastatic disease after 5 years in male patients with malignant melanomas who smoke, than in those who are non-smokers (Shaw et al., 1979). It has been generally accepted that one of the possible associations between smoking and the increased incidence of malignancy is related to the effects of smoking on the immune system. Ferson et al., (1979) first found the association between the low NK activity and smoking in human subjects by investigating age- and sex-matched smoking and non-smoking normal subjects and male, smoking and non-smoking melanoma patients. After that, there have been many reports confirming that cigarette smoking significantly inhibits human NK activity in peripheral blood lymphocytes (Ginns et al., 1985; Hughes et al., 1985; Phillips et al., 1985; Johnson et al., 1990; Jezewska et al., 1990; Zeidel et al., 2002; Sakami et al., 2002-03) and in bronchoalveolar lavage fluid (BALF) (Takeuchi et al., 1988; 2001) as an independent factor, or as a cofactor for the reduction of NK activity in patients of major depression (Jung and Irwin, 1999), suggesting an interaction between depression and cigarette smoking. We also found that cigarette smoking is associated with the reduction of NK activity (Kusaka et al., 1992; Inoue, et al., 1996b; Morimoto et al., 2001; 2005). Moreover, Hersey et al (1983) and Meliska et al., (1995) found that cessation of smoking increased NK activity after 1-3 months, whereas, the NK activity of smokers who continued to smoke did not show significant changes, providing direct evidence for this relationship. In addition, the findings from the report of Nakachi and Imai (1992) are noteworthy, in which cigarette smoking significantly reduced NK activity in a cross-sectional analysis of 2892 Japanese individuals. On the other hand, there have also been several studies reporting the negative effect of cigarette smoking on human NK activity (Nagel et al., 1981: Fulton et al., 1984; Fujita et al., 1985; Azari et al., 1996; Yovel et al., 2000; Koga et al., 2001; Nakamura et al., 1999; 2003;). Newman et al., (1991) even reported that NK activity was significantly higher in cigarette smokers compared with former smokers and control subjects that never smoked. Nagel et al., (1981) indicated that other factors such as age might obscure differences in NK activity based on smoking habits.

Alcohol consumption is another important lifestyle factor affecting human immune surveillance. There have been many papers reporting that NK activity in alcoholics is significantly lower than that in control subjects (Motivala SJ et al., 2003; Laso et al., 1997; Irwin et al., 1990; Redwine et al., 2003), suggesting that excessively chronic alcohol consumption could significantly inhibit human NK activity. Whereas, Kronfol et al., (1993) investigated NK activity in 47 hospitalized chronic alcoholic patients and 47 age- and sex-matched normal controls and found that there were no significant differences in NK activity between the two groups. Watzl et al (2002, 2004) reported that daily moderate amounts of red wine or alcohol have no effect on NK activity of healthy men in analyses of acute (2002) and two-week (2004) intakes comparing with dealcoholized red wine and red grape juice. Bounds et al (1994) also reported that social drinking does not affect human NK activity. Ochshorn-Adelson et al., (1994) found that a single dose of ethanol (0.5 g/kg), administered either intravenously (with resultant peak blood levels transiently up to 89 mg/dl) or orally (with resultant peak blood levels transiently up to 40 mg/dl at the time of the NK assay), did not alter NK activity. However, the results of in vitro studies showed a significant dose-dependent decrease ($p < 0.001$) in NK activity when ethanol exposure was sustained for 4 hr at concentrations of 80 mg/dl and above. Nakachi and Imai, (1992) even reported that alcohol drinking significantly increased human NK activity in a cross-sectional analysis of 2892 Japanese individuals.

Physical exercise is also an important lifestyle factor affecting human NK activity (Gleeson and Bishop, 2005; Nieman, 2000). It is generally accepted that light/moderate sport exercises enhance human NK activity (Nieman, 2000; Crist et al., 1989); trained athletes and people who have exercise habits show higher NK activity (Pedersen et al., 1989; Nieman, 2000; Levy et al., 1989; Kusaka et al., 1992), however, acute/heavy sports exercises induce transit decreases in NK activity (Nieman, 2000; Suzui et al., 2004). Brahmi et al., (1985) investigated the effect of acute exercise on NK activity in trained and sedentary individuals who underwent a standard progressive exercise test on a cycle ergometer, and found that NK activity against K562 reached maximum levels immediately after exercise, dropped to a low point 120 min later, then slowly came back to preexercise levels within 20 hr. However, no significant differences were observed between the trained and sedentary groups.

Mental stress (Levy et al., 1989; Evans et al., 1989; Kiecolt-Glaser et al., 1986; Kusaka et al., 1992; Borella et al., 1999; Koga et al., 2001) and nutrition condition (Abdallah et al., 1983; Nakachi and Imai, 1992; Venkatraman and Pendergast, 2002; Morimoto et al., 2004; Ishikawa et al., 2006) also affect human NK activity.

On the other hand, as described above, although there are many papers reporting the association between human NK activity and limited, single lifestyle factors, there have been few reports dealing with the association between human NK activity and overall lifestyle except for the studies reported by us (Kusaka et al., 1992, Inoue-Sakurai et al., 2000, Morimoto et al., 2001). To investigate the association of individual lifestyle with NK activity, we first assayed peripheral blood lymphocytes of 62 healthy males ranging in age from 30 to 60 years for NK cell activity, which was determined by ^{51}Cr release assay (Kusaka et al., 1992). Subjects were classified into groups reporting good, moderate, and poor lifestyles according to their responses on a questionnaire regarding eight health practices (tobacco smoking, alcohol consumption, hours of sleep, physical exercise, eating breakfast, balanced nutrition, hours of work, and mental stress) (Table 1). Subjects were asked about smoking status and were classified into three categories: current smoker, ex-smoker, and nonsmoker. Current smokers and ex-smokers were asked about the number of cigarettes smoked each day and about how many years they had been smoking. Alcohol drinking habits were assessed by asking about drinking frequency and alcohol consumption per occasion. Eating breakfast habits were classified into three categories: almost everyday, sometimes, or not. The physical exercise of the subjects was assessed by asking them about the frequency of doing physical exercise. Nutritional balance was classified into three categories: eating a balanced nutritional diet, eating with a little attention to a balanced diet, or do not eat a balanced diet, which was evaluated by self-perception. Eating a balanced nutritional diet was defined as follows: subjects take various vegetables and fruits, meat/fishes with a good balance in their daily life. Mental stress was also classified into three categories: feel excessive stress, mild, or slight stress, which was evaluated by self-perception. The score of these eight lifestyle factors was estimated according to Morimoto's criteria and are shown in Table 1. Each subject was then assigned a total score between 0 and 8 based on the total number of good health practices in Table 1 and classified into one of the following three categories using Morimoto's criteria: "poor lifestyle" (score=0-4), "moderate lifestyle" (score=5-6), "good lifestyle" (score=7-8). This method of classification has been verified as useful for the evaluation of personal lifestyle among Japanese (Hagihara and Morimoto, 1991; Kusaka et al., 1992; Inoue et al., 1996; Morimoto et al., 2001; Li et al., 2007a). Subjects reporting good lifestyle habits have a significantly higher NK activity than that in subjects reporting poor lifestyle habits. Among the eight health practices, cigarette smoking and physical exercise significantly affect NK activity. We further found that intake of fermented milk containing lactic acid bacteria was considered effective for restoring the NK cell activity of habitual smokers (Morimoto et al., 2004), and that aged garlic extract prevents a decline of NK activity in patients with advanced cancer (Ishikawa et al., 2006).

On January 17, 1995, at 5:46 AM, an earthquake with a magnitude of 7.2 on the Richter scale struck the Hanshin-Awaji district in the southern part of Hyogo Prefecture in Japan. About 6300 people were killed, more than 43000 people were injured, and over 436000 homes were damaged or destroyed. The central area of Kobe city was almost completely destroyed. This earthquake wreaked more havoc than any earthquake in Japan for the past several decades (Inoue-Sakurai et al., 2000). It has been reported that victims in the Hanshin-Awaji earthquake showed symptoms of posttraumatic stress disorder (PTSD) during the 3rd to 8th weeks after the earthquake (Kato et al., 1996). To investigate whether the lifestyle after the earthquake affected human NK activity, about 1.3 years after the earthquake, 155 male workers who experienced the event were administered questionnaires concerning their mental status, such as symptoms of PTSD, lifestyle, and demographic variables. Subjects who had PTSD symptoms showed lower NK cell activity than those without symptoms. Subjects with positive lifestyles showed higher NK cell activity than those with poor or moderate lifestyles. When subjects were divided into four groups by lifestyle and PTSD symptoms, subjects with positive lifestyles and few or no PTSD symptoms showed the highest NK cell activity among the four groups. The other three groups were subjects with positive lifestyles but many PTSD symptoms; subjects with poor or moderate lifestyles and many PTSD symptoms; and subjects with poor or moderate lifestyles and few or no PTSD symptoms. Taken together, PTSD symptoms and lifestyle were associated with NK cell activity in the earthquake victims (Inoue-Sakurai et al., 2000). Other reports have also pointed out the effect of PTSD on human NK activity (Delahanty et al., 1997; Laudenslager et al., 1998; Kawamura et al., 2001).

Effect of Lifestyle on LAK Activity

Lymphokine-activated killer (LAK) cells are subpopulations of lymphocytes that seem to play important roles in the immune response against cancers. LAK cells are generated when lymphocytes are stimulated by interleukin-2, which is released from T cells. LAK cells can kill a wide variety of target tumor cells including NK-cell-resistant cell lines and autologous and allogeneic tumor cells (Grimm et al., 1982; Allavena et al., 1987; Gambacorti-Passerini et al., 1988). Immunotherapy using LAK cells is a candidate treatment modality that may be effective in patients with tumors that are resistant to chemotherapy in breast, ovarian (Savas et al., 1996), and colon carcinomas (Rivoltini et al., 1991), small cell lung cancer (Savas et al., 1998), and melanoma (Savas et al., 1999). It is very important to study the association between lifestyle and LAK cell activity. However, the effects of lifestyle factors on human LAK cell activity are not well known. Lindemann and Park (1988) showed that smokeless tobacco extracts, when cultured with peripheral blood lymphocytes, inhibit the development of LAK cells. We have determined the LAK activity in a population consisting of 54 healthy Japanese males, including 23 smokers, 8 ex-smokers, and 23 non-smokers and found that cigarette smoking is associated with the reduction of LAK activity (Inoue, et al., 1996b). On the other hand, Tuschl et al. (1997) reported that there was no significant difference in LAK activity between smokers and non-smokers. Tvede et al. (1993) and Ullum et al (1994) reported that light, moderate, and intense acute bicycle exercise or acute bicycle ergometer exercise increased LAK activity. Whereas, Rohde et al., (1998) reported that in relation to intense exercise the lymphocyte concentration, the proliferative response, the NK and LAK cell activity declined. Laso et al., (1997) reported that decreased NK cytotoxic activity in chronic alcoholism is associated with alcohol liver disease. Moreover, Nair et al., (1990) reported that the generation and lytic capacity of LAK cells was significantly depressed by ethanol in vitro. Whereas, Kronfol et al. (1993) did not find significant differences in LAK activity between normal and alcoholic patients. Taken together, there are many inconsistencies between lifestyle factors and human LAK activity among current reports. Thus, we determined whether overall lifestyle affects human LAK cell activity by ^{51}Cr release assay in 54 healthy males. Subjects were classified into two groups, good or poor overall lifestyle, according to their answers on a questionnaire regarding eight health practices (cigarette smoking, alcohol consumption, physical exercise, mental stress, sleeping pattern, nutritional balance, eating breakfast, and working pattern). Subjects with good health practices with regard to smoking showed significantly higher LAK cell activity. LAK cells activity was significantly higher in subjects with a good overall lifestyle than in those with a poor overall lifestyle after controlling for the effects of smoking suggesting that lifestyle factors are associated with LAK cell activity (Inoue et al., 1996ab, Morimoto et al., 2001).

Effect of Lifestyle on Perforin, Granzyme and Granulysin: Mechanism of Lifestyle-Induced Inhibition of NK Activity

We found previously that overall lifestyle significantly affects human NK activity (Kusaka, et al., 1992; Inoue-Sakurai et al., 2000; Morimoto et al, 2001; Morimoto et al, 2005; Ishikawa et al., 2006) and LAK activity (Inoue et al., 1996ab; Morimoto et al, 2001), in which people with good lifestyles have higher NK and LAK activities than people with poor lifestyles. However, there have been no reports exploring the mechanism of decreased NK and LAK activities in people with poor lifestyles from the aspect of the NK killing mechanism.

NK and LAK cells induce tumor or virus infected target cell death by two main mechanisms (Kagi et al., 1994a; Lowin et al., 1994; Smyth et al., 2001). One mechanism involves granule exocytosis, with the direct release of cytolytic granules containing perforin, granzymes (Shinkai et al., 1988; Beresford et al., 1997; Smyth et al., 2001), and granulysin (GRN) (Jongstra et al., 1987; Okada et al., 2003; Krensky and Clayberger, 2005) that kill target cells via apoptosis. A second mechanism involves receptor-ligand interactions between Fas and Fas ligand (Kagi et al., 1994ab; Li et al., 2004). Cytotoxicity mediated by NK cells is greatly impaired in perforin-deficient mice (Kagi et al., 1994b; Li et al., 2004). Granzyme A (GrA) plays a critical role in triggering apoptosis in target cells either directly or via the activation of cellular caspases, and also cleaves IL-1β, a nucleosome assembly protein called putative HLA-associated protein II, TAF-Iβ, histones, and lamins (Beresford et al., 1997; Zhang et al., 2001a; Zhang et al., 2001b). Granzyme (GrB) directly cleaves downstream caspase substrates, nuclear matrix antigen, catalytic subunit of DNA-associated DNase inhibitor, and lamins (Zhang et al., 2001a; Smyth et al., 2001). In GrAxB-/- mice, NK-mediated target cell DNA fragmentation was not observed, even after extended incubation periods (10 h), but was normal in GrA-deficient and only impaired in GrB-deficient mice in short-term (2-4 h), but not long-term (4-10 h), nucleolytic assays. This suggests that GrA and GrB are critical for NK granule- mediated nucleolysis, with GrB being the main contributor (Simon et al., 1997). GRN, a lytic molecule expressed by human CTL and NK cells, is active against tumor cells and a variety of microbes. GRN can enter target cells in the absence of perforin and induce apoptosis, although GRN and perforin together are required to kill intracellular microbes like *Mycobacteria tuberculosis* (Okada et al., 2003; Stenger et al., 1998). Granulysin is associated with diverse activities of NK cells and CTL in physiological and pathological settings and could be a useful novel serum marker to evaluate the overall status of host cellular immunity (Ogawa et al., 2003).

To explore the mechanism of the effect of lifestyle on human NK activity, we investigated whether lifestyle affects the expression of perforin, granzymes, and GRN in NK cells and the levels of NK and T cells (Li et al., 2007a). In this study, 114 healthy male subjects aged 20-59 years from workers employed at a metal-working tool manufacturing corporation with a workforce of 350 workers in Osaka, Japan were selected with informed consent. Some of the factory workers had the possibility of exposure to various metal dusts, such as tungsten, titanium, cobalt, and carbon; however, there is a local ventilation system for their grinding machines in the factory, and the concentrations of these metal dusts in the workplace air were kept at very low concentrations (under the occupational exposure limits of the Japan Society for Occupational Health). These chemicals were measured every six month according to Japanese law for occupational health. In addition, workers used protective equipment such as protective dust masks and protective gloves. Moreover, there was no exposure to organic solvents in this workplace. Therefore, the influence of exposure to harmful substances in the workplace on the immune function of workers can almost be ignored. We chose only males from 20 to 59 years old to avoid any possible effects of gender. The subjects were selected randomly from the workers who received a medical examination performed at their work place in July 2005, and written informed consent was obtained from all subjects. None of the subjects had any signs or symptoms of infectious disease, used drugs that might affect immunological analysis, or were taking any medications at the time of the study. Lifestyle habits were assessed by using a self-administered questionnaire as described in the previous section, and evaluated according to Morimoto's criteria as shown in Table 1. Each subject was then assigned a total score between 0 and 8 and classified into one of the three categories using Morimoto's criteria as described above. Peripheral blood was taken and the numbers of NK and T cells and perforin-, GRN-, and GrA/B-expressing cells in PBL were measured by flow cytometry. As a result, subjects with good or moderate lifestyles showed significantly higher numbers of NK cells and perforin-, GRN-, and GrA/B-expressing cells (Fig. 1) (Li et al., 2007a) and a significant lower number of T cells in PBL than subjects with poor lifestyles.

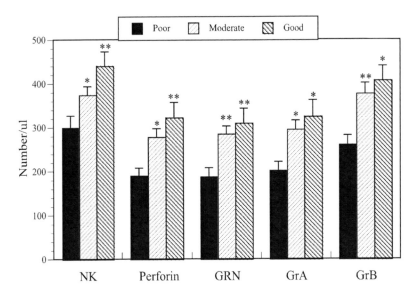

Figure 1. Effect of overall lifestyle on the numbers of $CD16^+$ NK cells, $perforin^+/CD16^+$, $GRN^+/CD16^+$, $GrA^+/CD16^+$, and $GrB^+/CD16^+$ cells in PBL. Data are presented as the mean +SE (n=114, except GrB, in which n= 70). ANOVA indicated that overall lifestyle significantly affected the numbers of $CD16^+$ NK cells, $perforin^+/CD16^+$, $GRN^+/CD16^+$, $GrA^+/CD16^+$, and $GrB^+/CD16^+$ cells (all p<0.05). *: p<0.05, **: p<0.01, significantly different from poor lifestyle by Tukey's method. Cited from Li et al., 2007a (Prev Med. 44, 117-123, 2007).

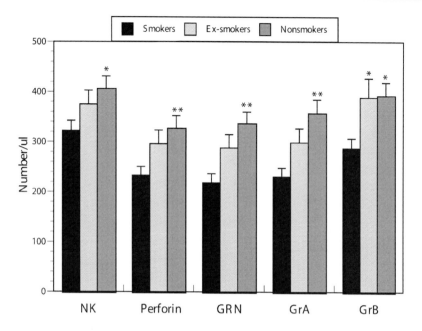

Figure 2. Effect of cigarette smoking on the numbers of CD16[+] NK cells, perforin[+]/CD16[+], GRN[+]/CD16[+], GrA[+]/CD16[+], and GrB[+]/CD16[+] cells in PBL. Data are presented as the mean +SE (n=114, except GrB, in which n= 70). ANOVA indicated that smoking significantly affected the numbers of CD16[+] NK cells, perforin[+]/CD16[+], GRN[+]/CD16[+], GrA[+]/CD16[+], and GrB[+]/CD16[+] cells (all $p<0.05$). *: $p<0.05$, **: $p<0.01$, significantly different from the smokers by Tukey's method. Cited from Li et al., 2007a (Prev Med. 44, 117-123, 2007).

We found that cigarette smoking significantly reduced the numbers of NK cells and perforin-, GRN-, and GrA /B-expressing cells (Fig. 2) (Li et al., 2007a). Other papers also reported that cigarette smoking significantly reduced the numbers of NK cells (Tollerud et al., 1989a; Inoue et al., 1996). On the other hand, Chrysofakis et al., (2004) reported that smoking increased the expression of perforin in sputum CD8[+] lymphocytes. However, there have been no reports on the effect of cigarette smoking on intracellular GRN or GrA/B in PBL although Glader et al., (2005) reported that cigarette smoke extract reduced the GrB-expressing cells in CD4[+], CD8[+] T cells *in vitro* after 72 h incubation. Interestingly, ex-smokers and non-smokers show almost the same levels in Gr/B-expressing cells and T cells, and ex-smokers show higher numbers of NK cell and perforin-, GRN-, and GrA-expressing cells than smokers although the differences were not significant, suggesting that the effect of smoking is reversible. This finding is very persuasive for anti-smoking education.

Concerning the effect of physical exercise on perforin, granzymes, and GRN, it has been reported that heavy exercise induces acute and transient decreases in perforin mRNA levels in lymphocytes (Miles et al., 2002) and perforin-positive cells in PBL (Staats et al., 2000). On the other hand, trained athletes and people who have exercise habits show higher NK cell counts and higher numbers of perforin- and GrB-expressing cells in lymphocytes (Staats et al., 2000; Kohut et al., 2004). We also found that subjects who have physical exercise habits show significantly higher numbers of NK cells, and perforin-, GRN-, and GrA/B-expressing cells. This is the first report proving the effect of physical exercise habits on GRN and GrA (Li et al., 2007a).

It has been reported that nutrition could affect human immunological functions including NK cell counts and NK activity (Amati et al., 2003; Albers et al., 2005). However, there have been no reports on the effect of nutrition on human perforin, GRN, GrA, or GrB so far. We first found that eating balanced nutritional diet showed a good effect on perforin and GrA in PBL, suggesting the importance of taking a balanced nutritional diet. The present findings also indirectly proved that nutritional condition may affect the intracellular perforin and GrA levels of NK cells (Li et al., 2007a)

It has been reported that alcohol consumption can affect human NK activity (Ochshorn-Adelson et al., 1994), the number of NK cells, and perforin expression in $CD3^+/CD56^+$ T cells and in NK cells, but does not affect the percentage of cells expressing granzyme (Perney et al., 2003). We found that drinking alcoholic beverages almost every day significantly reduced the number of GrA-expressing cells in NK cells, although the number of NK and T cells, and perforin-, GRN- and GrB-expressing cells was not affected, suggesting that alcohol drinking has a slight adverse effect on immune function (Li et al., 2007a).

Since $CD3^+$ T cells ($CD8^+$ cytotoxic cells) also express perforin, GRN, and GrA/B, we also examined the numbers of $perforin^+/CD3^+$, $GRN^+/CD3^+$, $GrA^+/CD3^+$, and $GrB^+/CD3^+$ cells; however, lifestyle factors did not affect these levels of these cells.

It has been reported that smokers showed higher proportions of T cells (Tollerud et al., 1989b; Inoue et al., 1996; Schaberg et al., 1997), and that mental stress increased T cell numbers in PBL (Liang et al., 1997; Bargellini et al., 2000). However, the effect of overall lifestyle on T cells in healthy subjects is not well known. We found that overall lifestyle, cigarette smoking, and eating breakfast significantly affected the number of T cells, and that a poor lifestyle significantly increased the number of T cells.

On the other hand, mental stress, working hours, and sleeping hours did not affect perforin-, GRN-, or GrA/B-expressing cells in PBL.

Taken together, the present study indicates that a poor lifestyle significantly decreases the numbers of NK cells and GRN-, perforin-, GrA/B-expressing cells in PBL and these decreases at least partially contribute to the decreased NK activity in the subjects with poor lifestyles that have been reported previously (Kusaka et al., 1992; Inoue et al., 1996b; Morimoto et al., 2001).

Effect of Forest Bathing (Natural Aromatherapy) on NK Activity, NK Number and Expression of Perforin, Granzyme, and Granulysin

A forest bathing trip (natural aromatherapy), called "Shinrinyoku" in Japanese, involves a visit to a forest field for the purpose of relaxation and recreation by breathing in volatile substances, called phytoncides (wood essential oils), which are antimicrobial volatile organic compounds derived from trees, such as alpha-pinene and limonene (Li et al. 2006, Li et al. 2007bc, Li et al., 2008ab). A forest bathing trip as a good lifestyle was first proposed in the 1980s and have become a recognized relaxation and/or stress management activity in Japan (Ohtsuka et al. 1998, Yamaguchi et al. 2006, Morita et al. 2007, Li et al. 2007bc, Li et al., 2008ab). The forest bathing trip significantly increased the score for vigor and decreased the scores for anxiety, depression, and anger as investigated by the Profile of Mood States

(POMS) test (Li et al. 2007b; 2008b). Customary forest bathing may help to decrease the risk of psychosocial stress-related diseases (Morita et al. 2007). Since forests occupy 67% of the land in Japan (Forestry Agency of Japan, 2002), forest bathing is easily accessible. According to a public opinion poll conducted in Japan in 2003, 25.6% of respondents had participated in a forest bathing trip, indicating its popularity in Japan (Morita et al. 2007). Moreover, forest bathing is possible in similar environments throughout the world. We reported previously that phytoncides enhance human NK activity and intracellular levels of perforin, granulysin, and granzyme A in NK cells *in vitro* (Li et al. 2006). Komori et al. (1995) also reported that citrus fragrance found in forests affects the human endocrine and immune systems as analyzed by the measurement of urinary cortisol and dopamine levels, NK activity, and CD4/8 ratios. These findings strongly suggest that forest bathing may have beneficial effects on human immune function.

Several studies have reported the adverse effect of long distance air travel on health, including the effect of jet lag (Waterhouse et al. 2004), health issues of air travel (DeHart, 2003), and economy class syndrome (Sahiar and Mohler 1994). In contrast, Toda et al (2006) reported that even short-period leisure trips (2 nights, 3 days) can bring about a reduction in di-stress and acquisition of eu-stress by analyzing the level of salivary endocrinological stress markers cortisol and chromogranin A. However, there have been no reports investigating the effect of short-period leisure trips on human NK function.

Thus, we investigated the effect of a forest bathing trip on human NK activity, NK cell numbers, and intracellular levels of perforin, GRN, and GrA/B in PBL (Li et al. 2007b). In this study, twelve healthy male subjects, aged 37-55 years (43.1 ± 6.1), were selected from three large companies in Tokyo, Japan. The information gathered from a self-administered questionnaire including age and lifestyle habits has been reported previously (Li et al., 2007a). None of the subjects had any signs or symptoms of infectious disease, used drugs that might affect immunological analysis, or were taking any medications at the time of the study.

The subjects experienced a three-day/two-night trip at three different forest fields in early September, 2005. On the first day, the subjects walked for two hours in the afternoon in a forest field, and then stayed at a nearby hotel within the forest. On the second day, subjects walked for 2 hours each in the morning and afternoon, in two different forest fields. Each course was 2.5 km, closely resembling normal physical activity for the subjects on normal working days. Daily physical activity of the subjects was monitored with a pedometer and the sleep in duration was measured with a piezo-electric accelerometer, Actiwatch(R) (Mini Mitter Co. Inc., Sunriver), worn on the wrist of the non-dominant arm. The validation studies were previously reported (Kawada et al., 2001; Benson et al., 2004). Blood was sampled on the second and third days and three days prior to the trip as a control. Since it has been reported that human NK cell activity shows circadian rhythms (Angeli, 1992), all samples were obtained at 8:00 am. All blood samples were placed in an ice/water box at 4°C and assays performed within four hours of the blood drawing. NK activity, proportions of NK and T cells, and GRN-, perforin-, and GrA/B-expressing cells in PBL were measured.

As results, almost all of the subjects (11/12) showed higher NK activity after the trip (about a 50% increase) compared with before. There were significant differences in NK activity both before and after the trip and between days 1 and 2 (Fig. 3). The forest bathing trip also significantly increased the numbers of NK cells (Fig. 3). In order to explore the mechanism of enhancement of NK activity by forest bathing, we investigated the effect of forest bathing on the intracellular levels of perforin, GRN, and GrA/B in PBL and found that the forest bathing trip also significantly increased the numbers of perforin-, GRN-, and GrA/B-expressing cells (Fig. 4). Taken together, these findings indicate that a forest bathing trip can increase NK activity, and that this effect at least partially mediated by increasing the number of NK cells and by the induction of intracellular anti-cancer proteins.

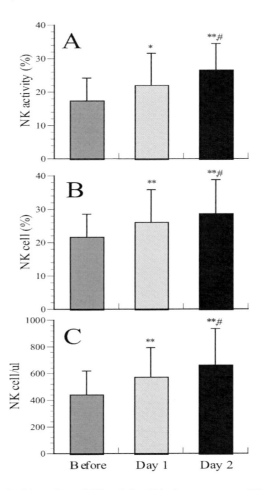

Figure 3. Effect of a forest bathing trip on NK activity (A), the percentage (B) and total number (C) of NK cells. Data are presented as the mean+SD (n=12). ANOVA indicated that the forest bathing trip significantly affected the NK activity, the percentage and total number of NK cells (all $p < 0.01$). *: $p < 0.05$, **: $p < 0.01$, significantly different from before the trip, #: $p < 0.05$ significantly different from Day 1 by paired t-test. The activity values for an E/T ratio of 20/1 are shown, and similar results were also obtained with E/T ratios of 40/1 and 10/1. Cited from Li et al., 2007b (Int. J. Immunopathol. Pharmacol. 20 (S2), 3-8, 2007).

Nevertheless, two questions remained to be resolved: (i) Will a trip to places without forest (a city tourist visit) also increase NK activity? (ii) How long does the increased NK activity last after a forest bathing trip or a city tourist visit? Thus, we have conducted two investigations to address these two questions (Li et al., 2008a). In the first investigation, twelve healthy male subjects, aged 35-56 years (mean 45.1±6.7), were selected from four large companies in Tokyo, Japan. Information gathered from a self-administered questionnaire, including age and lifestyle habits as described previously (Li et al., 2007b). None of the subjects had any signs or symptoms of infectious disease, used drugs that might affect immunological analysis, or were taking any medications at the time of the study. The subjects experienced a three-day/two-night trip to three different forest fields at Agematsu town in Nagano prefecture located in northwestern Japan in early September, 2006. The schedule of the forest bathing trip was similar to that described previously (Li et al., 2007b). In contrast, in the city tourist visit, eleven subjects experienced a three-day/two-night trip to Nagoya city in mid-May, 2006. Nagoya is located in Aichi prefecture roughly in the center of Japan with the population of 2,219,515, and is an important crossroads for transportation in Japan. On the first day, the subjects walked for two hours in the afternoon along a tourist route, which is through an old style district in Nagoya, and then stayed at a hotel in Nagoya. On the second day, the subjects walked for 2 hours around Nagoya baseball Dome in the morning and 2 hours around/in Nagoya airport nearby Nagoya city in the afternoon. There are some areas of trees in Nagoya city, but there are almost no trees in the areas visited. The class of hotel was the same and the hotel experience was the same for the city and the forest trips. The distance between Tokyo and the forest fields at Agematsu town in Nagano prefecture is almost the same as the distance between Tokyo and Nagoya. Each course in both trips was 2.5 km, closely resembling normal physical activity for the subjects on an average working day. Daily physical activity of the subjects was monitored with a pedometer, and the level of background walking steps of the subjects on normal working days were monitored for a week, and the averaged walking distance of all the subjects was about 5.0 km/day. Then, we set the walking distances in both trips based on the result. Blood was sampled on the second and third days during the trips, and on days 7 and 30 after the forest bathing trip, and three days prior to the trips as a control. White blood cell (WBC) counts, NK activity, proportions of NK and T cells, and GRN-, perforin-, and GrA/B-expressing cells in PBL were measured. Adrenaline concentration in urine was also determined. The forest bathing trip significantly increased human NK activity, the number of CD16$^+$ NK cells, and the percentages of GRN-, perforin-, and GrA/B-expressing cells in PBL, which confirmed the previous findings (Li et al., 2007b). The increased NK activity, number of CD16$^+$ NK cells, and the percentages of GRN-, perforin-, and GrA/B-expressing cells lasted more than 7 days, even 30 days. In contrast, the city tourist visit did not increase human NK activity, numbers of NK cells, nor the expression of the selected intracellular anti-cancer proteins. Moreover, the rates of increased NK activity and the number of NK cells during the forest bathing trip were significantly greater than those during the city tourist visit. The forest bathing trip did not affect lymphocyte or WBC counts. Neither the forest bathing trip nor the city tourist visit affected T cell numbers (Li et al., 2008a).

Moreover, we also found that a forest bathing trip significantly increased human NK activity, the number of CD16$^+$ NK cells, and the percentages of GRN-, perforin-, and GrA/B-expressing cells in PBL, and that the increased NK activity, number of CD16$^+$ NK cells, and the percentages of GRN-, perforin-, and GrA/B-expressing cells lasted more than 7 days in healthy female subjects (nurses) (Li et al., 2008b).

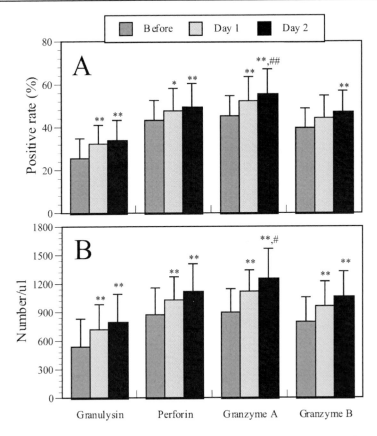

Figure 4. Effect of a forest bathing trip on the proportion (A) and number (B) of GRN-, perforin-, GrA/B-expressing cells in PBL. Data are presented as the mean +SD (n=12). ANOVA indicated that the forest bathing trip significantly affected the proportion and number of GRN-, perforin-, GrA/B-expressing cells in PBL (all p<0.01). *: p<0.05, **: p<0.01, significantly different from before the trip, #: p<0.05, ##: p<0.01 significantly different from Day 1 by the paired t-test. Cited from Li et al., 2007b (Int. J. Immunopathol. Pharmacol. 20 (S2), 3-8, 2007).

Adrenaline is released from the adrenal medulla, and the adrenaline level increases under circumstances of novelty, anticipation, unpredictability, and general emotional arousal, whereas the level of noradrenaline increases during increased physical activity (Frankenhaeuser 1975). Measurement of free adrenaline in urine provides a reliable measure of the circulating concentration of adrenaline in the bloodstream and thus a measure of adrenal medulla activity (Moleman et al., 1992). The concentration of adrenaline in urine has been used to evaluate work related stress in lorry drivers (van der Beek et al., 1995), long distance coach drivers (Sluiter et al., 1998), and psychosocial stress (Dimsdale and Moss., 1980). We found that a forest bathing trip significantly decreased the adrenaline concentration in urine, while a city tourist visit had no effect, suggesting that the subjects were under conditions of lower stress during the forest bathing trip (Li et al., 2008ab). It has been reported that adrenaline inhibits human NK activity (Garland et al. 2003). We found previously that physical and/or psychological stress decreased NK activity, NK receptor levels, and mRNA transcription of granzymes and perforin in mice (Li et al. 2005b). The increased NK activity in a forest bathing trip may be related to an attenuated stress hormone response (adrenaline) associated with the forest bathing trip. In addition, the forest bathing trip significantly increased the score for vigor and decreased the scores for anxiety, depression, and anger in the Profile of Mood States (POMS) test, suggesting that the subjects were physiologically relaxed during the forest bathing trip (Li et al., 2007b). Moreover, we found that forest bathing significantly increased the proportions of lymphocytes and monocytes and decreased the proportions of granulocytes in WBC (Li et al., 2007b). It has been reported that dominance by the parasympathetic nervous system causes an increase in circulating lymphocytes and decrease in granulocytes in peripheral blood (Mori et al., 2002). This suggests that the parasympathetic nervous system of subjects was dominant, associated with relaxation and decreased stress during the forest bathing trips. Previous studies have reported that forest bathing reduces the concentration of cortisol in saliva, reduces prefrontal cerebral activity, reduces blood pressure, and stabilizes autonomic nervous activity in humans (Park et al. 2007, Yamaguchi et al. 2006).

Many factors, including circadian variation (Angeli 1992, Gamaleia 2006), physical exercise (Nieman 2000, Miles et al. 2002, Li et al. 2007a), and alcohol consumption (Ochshorn-Adelson et al. 1994, Li et al. 2007a) can affect human NK activity. In order to control the effect of circadian rhythm on NK activity, we sampled blood at 8:00 am on all days. To control for the effect of physical exercise on NK activity, we limited the walking steps during the trips to the averaged normal workday distances as monitored by a pedometer. The levels of physical activity between the two trips were also matched. To control the effect of alcohol on NK activity, the subjects did not consume alcohol during the study period for both trips. The sleeping hours during the trips were a little longer than on average working days; however, the difference was not significant in either type of trip. There are several reports addressing the effect of sleeping hours on NK cell activity. Many reports suggest that sleep deprivation increases human NK activity (Dinges et al. 1994; Matsumoto et al. 2001), while others suggest that sleep deprivation decreases human NK activity (Moldofsky et al. 1989, Irwin et al. 1994). Another study by Kusaka et al. (1992) reported that sleeping hours did not affect NK activity, or NK cell numbers under physiological conditions. We also found that there was no difference in the numbers of NK cells, nor the levels of perforin-, GRN-, or GrA/B-expressing cells in PBL among the subjects who slept 5, 6, or 7 hours (Li et al. 2007a). In addition, although the sleeping hours during the city tourist visit were a little longer than on average working days, the NK activities during the trip were almost the same as for working days, indicating that the longer sleeping hours did not contribute to NK activity in the city tourist visit. Taken together, although the sleeping hours during the trips were a little longer than those on average working days, this difference did not affect either NK activity or cell numbers in the present study.

Phytoncides, such as alpha-pinene and beta-pinene, were detected in forest air, but were almost absent in city air (Li et al., 2007b; 2008ab). These findings indicate that a forest bathing trip increased NK activity, the number of NK cells, and the levels of intracellular anti-cancer proteins, and that this effect lasted at least 7 days after the trip. Phytoncides released from trees and decreased stress hormone may partially contribute to the increased NK activity (Li et al., 2006).

Effect of Lifestyle and Forest Bathing on NKT Cells

Although it has been reported that lifestyles including forest bathing trips significantly affect human NK and LAK activities, NK cell number, and the intracellular perforin, GrA, GrB, and GRN levels (Kusaka, et al., 1992; Inoue et al., 1996ab; Inoue-Sakurai et al., 2000; Morimoto et al, 2001; Morimoto et al, 2005; Ishikawa et al., 2006; Li et al., 2007abc), there have been no reports exploring the effect of lifestyle and forest bathing on human NKT cells so far. NKT cells were originally named because of the co-expression of a T-cell receptor (TCR) along with typical surface receptors for NK cells including CD16, CD56, CD161, CD94, CD158a, and CD158b (Godfrey et al., 2000; Hamzaoui et al., 2006). Two subtypes of CD1d-restricted NKT cells have been identified. Type I NKT cells, also called invariant NKT (iNKT) cells, have a highly restricted T-cell receptor (TCR) repertoire and express TCR Vα14–Jα18 and Vβ8.2, -7 or -2 chains in mice and homologous TCR Vα24–Jα18 and Vβ11 chains in humans. Type II NKT cells, which are also referred to as non-iNKT cells, express a more diverse TCR repertoire (Van Kaer, 2007). The $CD56^+/CD3^+$ NKT cells are also called NKT-like cells (Aguilar et al., 2006). NKT cells have been implicated in immune responses against infectious agents, tumors and tissue grafts and in regulating a variety of autoimmune, adaptive immune response, and inflammatory diseases (Godfrey et al., 2000; Van Kaer, 2007). In the present study, we first investigated whether lifestyle and forest bathing affect human $CD56^+/CD3^+$ NKT cells. In the lifestyle study, 70 healthy male subjects aged 20-59 years from a large company in Osaka, Japan were selected with informed consent. The subjects were divided into groups reporting good, moderate, and poor lifestyles according to their responses on a questionnaire regarding the same eight health practices as described previously (Li et al., 2007a). Blood was sampled and $CD56^+/CD3^+$ NKT cells were determined by flow cytometry. In the forest bathing study, twelve healthy male subjects, aged 35-56 years (mean 45.1±6.7), were selected from four large companies in Tokyo, Japan as described previously (Li et al. 2008a). The subjects experienced a three-day/two-night trip to three different forest fields at Agematsu town in Nagano prefecture located in northwest Japan in early September, 2006 (Li et al. 2008a). In contrast, in the city tourist visit, eleven subjects experienced a three-day/two-night trip to Nagoya city in mid-May, 2006 (Li et al. 2008a). Blood was sampled on the second and third days during the trips, and on days 7 and 30 after the forest bathing trip, and three days prior to the trips as a control. $CD56^+/CD3^+$ NKT cells were determined by flow cytometry.

Effect of Cigarette Smoking on $CD56^+/CD3^+$ NKT Cells

As shown in Fig. 5, nonsmokers showed a significantly higher percentage of $CD56^+/CD3^+$ NKT cells than smokers (p<0.05). Ex-smokers also showed a higher percentage of $CD56^+/CD3^+$ NKT cells than smokers, but the difference was not significant. Moreover, we found a significant correlation between the percentage of $CD56^+/CD3^+$ NKT cells and numbers of cigarettes smoked per day (r=-0.287, n=69, p<0.05) or the number of years the subject had smoked for (r=-0.330, n=69, p<0.01) in smokers. We further used the Brinkman index (numbers of cigarettes smoked per day X years smoked) to quantitatively measure the effect of smoking on $CD56^+/CD3^+$ NKT cells, and found a significant correlation between the Brinkman index and percentage of $CD56^+/CD3^+$ NKT cells (r=-0.304, n=69, p<0.05), indicating that cigarette smoking indeed reduced the percentage of CD56$^+$/CD3$^+$ NKT cells. On the other hand, there was no significant difference between nonsmokers and ex-smokers in terms of CD56$^+$/CD3$^+$ NKT cells (Li et al. 2007c).

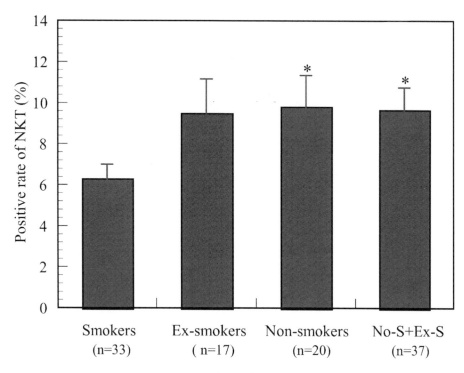

Figure 5. Effect of cigarette smoking on CD56$^+$/CD3$^+$ NKT cells in PBL. Data are presented as the mean +SE. *: p<0.05, significantly different from smokers by unpaired t-test.

Effect of Alcohol Drinking on CD56$^+$/ CD3$^+$ NKT Cells

As shown in Fig. 6, people who drink alcoholic beverages almost every day showed a significantly lower percentage of CD56$^+$/CD3$^+$ NKT cells than people who drink sometimes (p<0.05), indicating that alcohol drinking has an adverse effect on CD56$^+$/CD3$^+$ NTK cells. Surprisingly, people who never drink alcoholic beverages also show a lower percentage of CD56$^+$/CD3$^+$ NKT cells than people who drink sometimes although the difference was not significant, suggesting that taking a moderate quantity of alcohol may have a beneficial effect on CD56$^+$/CD3$^+$ NTK cells (Li et al. 2007c). There was only one paper reporting that CD8$^+$/CD56$^+$ T cells was lower in alcoholic patients than that in controls; however, the difference was not significant (Perney et al., 2003).

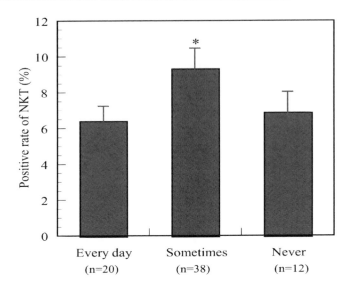

Figure 6. Effect of alcohol consumption on CD56$^+$/CD3$^+$ NKT cells in PBL. Data are presented as the mean +SE. *: p<0.05, significantly different from the subjects who drink alcohol every day by unpaired t-test.

Effect of Working Hours on NKT Cells

As shown in Fig. 7, subjects who worked 9 h every day showed a significantly lower percentage of CD56$^+$/CD3$^+$ NKT cells than subjects who worked 8 h every day (p<0.05), suggesting that longer working hours have an adverse effect on CD56$^+$/CD3$^+$ NTK cells (Li et al. 2007c).

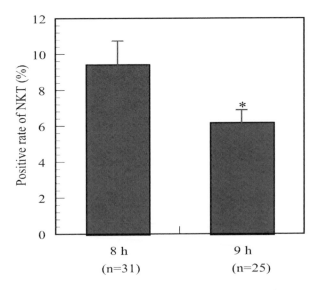

Figure 7. Effect of working hours on CD56$^+$/CD3$^+$ NKT cells in PBL. Data are presented as the mean +SE. *: p<0.05, significantly different from the subjects who work for 8 h every day by unpaired t-test.

Effect of Eating Balanced Nutritional Diet on CD56⁺/CD3⁺ NKT Cells

As shown in Fig. 8, subjects who ate a balanced nutritional diet every day showed a higher percentage of CD56$^+$/CD3$^+$ NKT cells than subjects who did not take balanced nutritional diet, although the difference was not significant (p=0.103) (Li et al. 2007c).

Mental stress, physical exercise, and sleeping hours did not affect CD56$^+$/CD3$^+$ NKT cells in PBL in the present study.

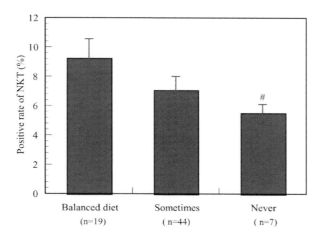

Figure 8. Effect of eating a balanced nutritional diet on CD56$^+$/CD3$^+$ NKT cells in PBL. Data are presented as the mean +SE. #: p=0.103, different from the subjects who eat a balanced diet by unpaired t-test.

Effect of Overall Lifestyle on CD56⁺/CD3⁺ NKT Cells

As shown in Fig. 9, the subjects with good or moderate lifestyles show a higher level of CD56$^+$/CD3$^+$ NKT cells in PBL compared with subjects with poor lifestyles although the difference was not significant (Li et al. 2007c).

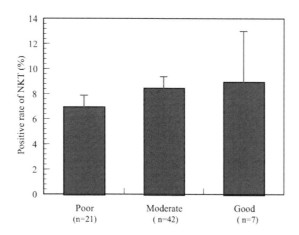

Figure 9. Effect of overall lifestyle on CD56$^+$/CD3$^+$ NKT cells in PBL. Data are presented as the mean +SE.

Effect of a Forest Bathing Trip Versus a City Tourist Visit on CD56⁺/CD3⁺ NKT Cells

The forest bathing trip significantly increased the percentage of CD56$^+$/CD3$^+$ NKT cells, and this increase lasted more than 7 days after the trip (Fig. 10A). In contrast, the city tourist visit did not increase human CD56$^+$/CD3$^+$ NKT cells (Fig. 10B). The forest bathing trip did not affect lymphocyte or WBC counts (Li et al. 2007c).

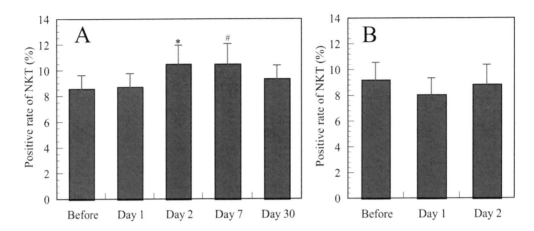

Figure 10. Effect of a forest bathing trip (A) *versus* a city tourist visit (B) on CD56$^+$/CD3$^+$ NKT cells in PBL. Data are presented as the mean+SE (n=12 in A and n=11 in B). Two-way ANOVA with no-repeated measures indicated that the forest bathing trip (p<0.05) and the variability between individuals (p<0.01) significantly affected the percentage of CD56$^+$/CD3$^+$ NKT cells. *: *p*<0.05, #: p=0.056 significantly different from before the trip by paired t-test.

Taken together, these findings indicate that a good lifestyle including forest bathing trips significantly increases the positive rate of human CD56$^+$/CD3$^+$ NKT cells in PBL.

Conclusion

Good lifestyles such as physical exercise, forest bathing trips, and healthy nutrition conditions have beneficial effects on human NK and LAK activities, whereas poor lifestyles such as cigarette smoking, excessive alcohol consumption, and mental stress have adverse effects on human NK and LAK activities.

Lifestyles affect human NK and LAK activities at least partially mediated by the effect on the numbers of NK and CD56$^+$/CD3$^+$ NKT cells, and perforin-, GRN-, and GrA/B-expressing cells.

Acknowledgements

This work was supported by a grant from the Ministry of Education, Culture, Sports, Science, and Technology of Japan (14207017) and a research project for utilizing advanced technologies in agriculture, forestry, and fisheries of Japan (2004-2006). We are grateful to

Professor Tomoyuki Kawada and other staff of the Department of Hygiene and Public Health, Nippon Medical School, for their advice and assistances.

References

Abdallah, RM; Starkey, JR; Meadows, GG. Alcohol and related dietary effects on mouse natural killer-cell activity. *Immunology.* 1983; 50, 131-7.

Aguilar, P; Mathieu, CP; Clerc, G; Ethevenot, G; Fajraoui, M; Mattei, S; Faure, GC; Bene, MC. Modulation of natural killer (NK) receptors on NK (CD3-/CD56+),T(CD3+/CD56-) and NKT-like (CD3+/CD56+) cells after heart transplantation. *J Heart Lung Transplant.* 2006; 25, 200-5.

Albers, R; Antoine, JM; Bourdet-Sicard, R; Calder, PC; Gleeson, M; Lesourd, B; Samartin, S; Sanderson, IR; Van Loo, J; Vas Dias, FW; Watzl, B. Markers to measure immunomodulation in human nutrition intervention studies. *Br. J. Nutr.* 2005; 94, 452-481.

Allavena, P; Grandi, M; D'Incalci, M; Geri, O; Giuliani, FC; Mantovani, A. Human tumor cell lines with pleiotropic drug resistance are efficiently killed by interleukin-2 activated killer cells and by activated monocytes. *Int J Cancer.* 1987; 40, 104–107.

Amati, L; Cirimele, D; Pugliese, V; Covelli, V; Resta, F; Jirillo, E. Nutrition and immunity: laboratory and clinical aspects. *Curr. Pharm. Des.* 2003; 9, 1924-1931.

Angeli, A. Circadian rhythms of human NK cell activity. *Chronobiologia* 1992; 19, 195-8.

Azari, MR; Williams, FM; Kirby, J; Kelly, P; Edwards, JW; Blain, PG. Effects of nitrogen oxides on natural killer cells in glass craftsmen and braziers. *Occup Environ Med.* 1996; 53, 248-51.

Bargellini, A; Barbieri, A; Rovesti, S; Vivoli, R; Roncaglia, R; Borella, P. Relation between immune variables and burnout in a sample of physicians. *Occup. Environ. Med.* 2000; 57, 453-457.

Benson, K; Friedman, L; Noda, A; Wicks, D; Wakabayashi, E; Yesavage, J. The measurement of sleep by actigraphy: direct comparison of 2 commercially available actigraphs in a nonclinical population. Sleep 2004; 27, 986-989.

Beresford, PJ; Kam, CM; Powers, JC; Lieberman, J. Recombinant human granzyme A binds to two putative HLA-associated proteins and cleaves one of them. *Proc. Natl. Acad. Sci. U S A.* 1997; 94, 9285-9290.

Berkman, LF; Breslow, L. Health and ways of living: the Alameda County Study. New York: Oxford Univ. Press; 1983.

Borella, P; Bargellini, A; Rovesti, S; Pinelli, M; Vivoli, R; Solfrini, V; Vivoli, G. Emotional stability, anxiety, and natural killer activity under examination stress. *Psychoneuroendocrinology.* 1999; 24, 613-27.

Bounds, W; Betzing, KW; Stewart, RM; Holcombe, RF. Social drinking and the immune response: impairment of lymphokine-activated killer activity. *Am J Med Sci.* 1994; 307, 391-5.

Brahmi, Z; Thomas, JE; Park, M; Park, M; Dowdeswell, IR. The effect of acute exercise on natural killer-cell activity of trained and sedentary human subjects. *J Clin Immunol.* 1985; 5, 321-8.

Chrysofakis, G; Tzanakis, N; Kyriakoy, D;, Tsoumakidou, M; Tsiligianni, I; Klimathianaki, M; Siafakas, NM. Perforin expression and cytotoxic activity of sputum CD8+ lymphocytes in patients with COPD. Chest 2004; 125, 71-76.

Crist, DM; Mackinnon, LT; Thompson, RF; Atterbom, HA; Egan, PA. Physical exercise increases natural cellular-mediated tumor cytotoxicity in elderly women. *Gerontology.* 1989; 35, 66-71.

DeHart, RL. Health issues of air travel. *Annu Rev Public Health* 2003; 24, 133-51.

Delahanty, DL; Dougall, AL; Craig, KJ; Jenkins, FJ; Baum, A. Chronic stress and natural killer cell activity after exposure to traumatic death. *Psychosom Med.* 1997; 59(5), 467-76.

Dinges, DF; Douglas, SD; Zaugg, L; Campbell, DE; McMann, JM; Whitehouse, WG.; Orne, EC; Kapoor, SC; Icaza, E; Orne, MT. Leukocytosis and natural killer cell function parallel neurobehavioral fatigue induced by 64 hours of sleep deprivation. *J. Clin. Invest.* 1994, 93; 1930-1939.

Dimsdale, JE; Moss, J. Plasma catecholamines in stress and exercise. *JAMA.* 1980; 243, 340-2.

Doll, R; Peto, R. Mortality in relation to smoking: 20 years' observations on male British doctors. *Br Med J.* 1976; 2(6051), 1525-36.

Evans, DL; Leserman, J; Pedersen, CA; Golden, RN; Lewis, MH; Folds, JA; Ozer H. Immune correlates of stress and depression. *Psychopharmacol Bull.* 1989; 25, 319-24.

Ferson, M; Edwards, A; Lind, A; Milton, GW; Hersey, P. Low natural killer-cell activity and immunoglobulin levels associated with smoking in human subjects. *Int J Cancer.* 1979; 23, 603-9.

Forestry Agency of Japan. http://www.rinya.maff.go.jp/toukei/genkyou/shinrin-jinkou.htm, 2002.

Frankenhaeuser, M., Experimental approach to the study of catecholamines and emotion. In: Levi L, editor. *Emotions, Their Parameters and Measurement.* New York: Raven Press. 1975; pp209-234.

Fujita, J; Saijo, N; Sasaki, Y; Sakurai, M; Sano, T; Ishihara, J; Takahashi, H; Hoshi, A. Natural killer (NK) and lymphokine-activated killer (LAK) cell activity in healthy volunteers with special emphasis on the familial incidence of cancer. *Jpn J Clin Oncol.* 1985; 15, 589-94.

Fulton, A; Heppner, G; Roi, L; Howard, L; Russo, J; Brennan, M. Relationship of natural killer cytotoxicity to clinical and biochemical parameters of primary human breast cancer. *Breast Cancer Res Treat.* 1984; 4, 109-16.

Gamaleia, NF; Skivka, LM; Fedorchuk, AG; Shishko, ED. Circadian rhythms of cytotoxic activity in peripheral blood mononuclear cells of patients with malignant melanoma. *Exp Oncol.* 2006; 28, 54-60.

Gambacorti-Passerini, C; Rivoltini, L; Supino, R; Rodolfo, M; Radrizzani, M; Fossati G; Parmiani, G. Susceptibility of chemoresistant murine and human tumor cells to lysis by interleukin 2-activated lymphocytes. *Cancer Res.* 1988; 48, 2372–2376.

Garland, M; Doherty, D; Golden-Mason, L; Fitzpatrick, P; Walsh, N; O'Farrelly, C. Stress-related hormonal suppression of natural killer activity does not show menstrual cycle variations: implications for timing of surgery for breast cancer. *Anticancer Res.* 2003; 23(3B), 2531-5.

Ginns, LC; Ryu, JH; Rogol, PR; Sprince, NL; Oliver, LC; Larsson, CJ. Natural killer cell activity in cigarette smokers and asbestos workers. *Am Rev Respir Dis.* 1985; 131, 831-4.

Glader, P; Moller, S; Lilja, J; Wieslander, E; Lofdahl, CG; von Wachenfeldt, K. Cigarette smoke extract modulates respiratory defence mechanisms through effects on T-cells and airway epithelial cells. *Respir. Med.* 2006; 100, 818-27.

Gleeson, M; Bishop, NC. The T cell and NK cell immune response to exercise. *Ann Transplant.* 2005; 10; 43-8.

Godfrey, DI; Hammond, KJ; Poulton, LD; Smyth, MJ; Baxter, AG. NKT cells: facts, functions and fallacies. *Immunol Today.* 2000; 21, 573-83.

Grimm, EA; Mazumder, A; Zhang, HZ; Rosenberg, SA. Lymphokine-activated killer cell phenomenon. Lysis of natural killer-resistant fresh solid tumor cells by interleukin 2-activated autologous human peripheral blood lymphocytes. *J Exp Med.* 1982; 155, 1823-41.

Hagihara, A; Morimoto, K. Personal health practices and attitudes toward nonsmokers' legal rights in Japan. *Soc. Sci. Med.* 1991; 33, 717-721.

Hamzaoui, A; Rouhou, SC; Grairi, H; Abid, H; Ammar, J; Chelbi, H,; Hamzaoui, K. NKT cells in the induced sputum of severe asthmatics. *Mediators Inflamm.* 2006; 2006, 71214, 1-6.

Hersey, P; Prendergast, D; Edwards, A. Effects of cigarette smoking on the immune system. Follow-up studies in normal subjects after cessation of smoking. *Med J Aust.* 1983; 2, 425-9.

Hirayama, T. Life style and mortality: Large scale census-based cohort study in Japan. In: Wahrendorf, J. (Ed.), *Contribution to Epidemiology and Biostatistics*. Basel: Karger, 1990, pp. 1-137.

Hughes, DA; Haslam, PL; Townsend, PJ; Turner-Warwick, M. Numerical and functional alterations in circulatory lymphocytes in cigarette smokers. *Clin Exp Immunol.* 1985; 61, 459-66.

Imai, K; Matsuyama, S; Miyake, S; Suga, K; Nakachi, K. Natural cytotoxic activity of peripheral-blood lymphocytes and cancer incidence: an 11-year follow-up study of a general population, *Lancet* 2000; 356, 1795–1799.

Inoue-Sakurai, C; Maruyama, S; Morimoto, K. Posttraumatic stress and lifestyles are associated with natural killer cell activity in victims of the Hanshin-Awaji earthquake in Japan. *Prev Med.* 2000; 31, 467-73.

Inoue, C; Takeshita, T; Kondo, H; Morimoto, K. Healthy lifestyles are associated with higher lymphokine-activated killer cell activity. *Prev. Med.* 1996; 25, 717-724.

Inoue, C; Takeshita, T; Kondo, H; Morimoto, K. Cigarette smoking is associated with the reduction of lymphokine-activated killer cell and natural killer activities. *Environmental Health and Preventive Medicine* 1996; 1, 14-19.

Irwin, M; Caldwell, C; Smith, TL; Brown, S; Schuckit, MA; Gillin, JC. Major depressive disorder, alcoholism, and reduced natural killer cell cytotoxicity. Role of severity of depressive symptoms and alcohol consumption. *Arch Gen Psychiatry.* 1990; 47, 713-9.

Irwin, M; Mascovich, A; Gillin, JC; Willoughby, R; Pike, J; Smith, TL. Partial sleep deprivation reduces natural killer cell activity in humans. Psychosom. Med. 51994; 6, 493-498.

Ishikawa, H; Saeki, T; Otani, T; Suzuki, T; Shimozuma, K; Nishino, H; Fukuda, S; Morimoto, K. Aged garlic extract prevents a decline of NK cell number and activity in patients with advanced cancer. *J Nutr.* 2006; 136(3 Suppl), 816S-820S.

Jezewska, E; Dworacki, G; Skrzypczak, A; Zeromski, J. Surface antigens and cytotoxic natural killer cell (NK) activity of blood lymphocytes in heavy cigarette smokers. *Arch Geschwulstforsch.* 1990; 60(3), 187-92.

Johnson, JD; Houchens, DP; Kluwe, WM; Craig, DK; Fisher, GL. Effects of mainstream and environmental tobacco smoke on the immune system in animals and humans: a review. *Crit Rev Toxicol.* 1990; 20, 369-95.

Jongstra, J; Schall, TJ; Dyer, BJ; Clayberger, C; Jorgensen, J; Davis, MM; Krensky, AM., The isolation and sequence of a novel gene from a human functional T cell line. *J. Exp. Med.* 1987; 165, 601-614.

Jung W, Irwin M. Reduction of natural killer cytotoxic activity in major depression: interaction between depression and cigarette smoking. Psychosom Med. 1999; 61, 263-70.

Kagi, D; Vignaux, F; Ledermann, B; Burki, K; Depraetere, V; Nagata, S; Hengartner, H; Golstein, P. Fas and perforin pathways as major mechanisms of T cell-mediated cytotoxicity. *Science* 1994a; 265, 528-530.

Kagi, D; Ledermann, B; Burki, K; Seiler, P; Odermatt, B; Olsen, KJ; Podack, ER; Zinkernagel, RM; Hengartner, H. Cytotoxicity mediated by T cells and natural killer cells is greatly impaired in perforin-deficient mice. *Nature* 1994b; 369, 31-37.

Kato, H; Asukai, N; Miyake, Y; Minakawa, K; Nishiyama, A. Post-traumatic symptoms among younger and elderly evacuees in the early stages following the 1995 Hanshin-Awaji earthquake in Japan. *Acta Psychiatr Scand* 1996; 93, 477–481.

Kawada, T; Xin, P; Kuroiwa, M; Sasazwa, Y; Suzuki, S; Tamura, Y. Habituation of sleep to road traffic noise as determined by polysomnography and accelerometer. *J. Sound Vib.* 2001; 242, 169-178.

Kawamura, N; Kim, Y; Asukai, N. Suppression of cellular immunity in men with a past history of posttraumatic stress disorder. *Am J Psychiatry.* 2001; 158, 484-6.

Kiecolt-Glaser, JK; Glaser, R; Strain, EC; Stout, JC; Tarr, KL; Holliday, JE; Speicher, CE. Modulation of cellular immunity in medical students. *J Behav Med.* 1986; 9, 5-21.

Koga, C; Itoh, K; Aoki, M; Suefuji, Y; Yoshida, M; Asosina, S; Esaki, K; Kameyama, T. Anxiety and pain suppress the natural killer cell activity in oral surgery outpatients. *Oral Surg Oral Med Oral Pathol Oral Radiol Endod.* 2001; 91; 654-8.

Kohut, ML; Arntson, BA; Lee, W; Rozeboom, K; Yoon, KJ; Cunnick, JE; McElhaney, J. Moderate exercise improves antibody response to influenza immunization in older adults. *Vaccine* 2004; 22, 2298-306.

Komori, T; Fujiwara, R; Tanida, M; Nomura, J; Yokoyama, MM. Effects of citrus fragrance on immune function and depressive states. *Neuroimmunomodulation* 1995; 2, 174-180.

Krensky, AM; Clayberger, C. Granulysin: a novel host defense molecule. *Am. J. Transplant.* 2005; 5, 1789-92.

Kronfol, Z; Nair, M; Hill, E; Kroll, P; Brower, K; Greden, J. Immune function in alcoholism: a controlled study. *Alcohol Clin Exp Res.* 1993; 17, 279-83.

Kusaka, Y; Kondou, H; Morimoto, K. Healthy lifestyles are associated with higher natural killer cell activity. *Prev. Med.* 1992; 21, 602-615.

Laso, FJ; Madruga, JI; Giron, JA; Lopez, A; Ciudad, J; San Miguel, JF; Alvarez-Mon, M; Orfao, A. Decreased natural killer cytotoxic activity in chronic alcoholism is associated with alcohol liver disease but not active ethanol consumption. *Hepatology.* 1997; 25, 1096-100.

Laudenslager, ML; Aasal, R; Adler, L; Berger, CL; Montgomery, PT; Sandberg, E; Wahlberg, LJ; Wilkins, RT; Zweig, L; Reite, ML. Elevated cytotoxicity in combat veterans with long-term post-traumatic stress disorder: preliminary observations. *Brain Behav Immun.* 1998; 12(1), 74-9.

Levy, SM; Herberman, RB; Simons, A. Persistently low natural killer cell activity in normal adults: Immunological, hormonal and mood correlates. *Nat. Immun. Cell Growth Regul.* 1989; 8, 173-186.

Li, Q; Nakadai, A; Takeda, K; Kawada, T. Dimethyl 2,2-dichlorovinyl phosphate (DDVP) markedly inhibits activities of natural killer cells, cytotoxic T lymphocytes and lymphokine-activated killer cells *via* the Fas-ligand/Fas pathway in perforin-knockout (PKO) mice. *Toxicology* 2004; 204, 41-50.

Li, Q; Liang, Z; Nakadai, A; Kawada, T. Effect of electric foot shock and psychological stress on NK, LAK and CTL activities, NK receptors and mRNA transcripts of granzymes and perforin. *Stress* 2005; 8, 107-116.

Li, Q; Nakadai, A; Matsushima, H; Miyazaki, Y; Krensky, AM; Kawada, T; Morimoto, K. Phytoncides (wood essential oils) induce human natural killer cell activity. *Immunopharmacol. Immunotoxicol.* 2006; 28, 319-33.

Li, Q; Morimoto, K; Nakadai, A; Qu, T; Matsushima, H; Katsumata, M; Shimizu, T; Inagaki, H; Hirata, Y; Hirata, K; Kawada, T; Lu, Y, Nakayama, K; Krensky, AM. Healthy lifestyles are associated with higher levels of perforin, granulysin and granzymes A/B - expressing cells in peripheral blood lymphocytes. *Prev Med.* 2007a; 44, 117-123.

Li, Q; Morimoto, K; Nakadai, A; Inagaki, H; Katsumata, M; Shimizu, T; Hirata, Y; Hirata, K; Suzuki, H; Miyazaki, Y; Kagawa, T; Koyama, Y; Ohira, T; Takayama, N; Krensky, AM; Kawada, T. Forest bathing enhances human natural killer activity and expression of anti-cancer proteins. *Int. J. Immunopathol. Pharmacol.* 2007b; 20 (S2), 3-8.

Li, Q; Morimoto, K; Kagawa, T. Lifestyle affects human NKT cells. *Proceedings of the 14th Annual Meeting of the Japanese Society of Immunotoxicology.* 2007c; pp43.

Li, Q; Morimoto, K; Kobayashi, M; Inagaki, H; Katsumata, M; Hirata, Y; Hirata, K; Suzuki, H; Li, YJ; Wakayama, Y; Kawada, T; Park, BJ; Ohira, T; Matsui, N; Kagawa, T; Miyazaki, Y; Krensky, AM. Visiting a forest, but not a city, increases human natural killer activity and expression of anti-cancer proteins. *Int. J. Immunopathol. Pharmacol.* 2008a; 21(1), 117-128.

Li Q, Morimoto K, Kobayashi M, Inagaki H, Katsumata M, Hirata Y, Hirata K, Suzuki H, Li YJ, Wakayama Y, Kawada T, Ohira T, Takayama N, Kagawa T, Miyazaki Y. A forest bathing trip increases human natural killer activity and expression of anti-cancer proteins in female subjects. J Biol Regul Homeost Agents 2008b (in press).

Liang, SW; Jemerin, JM; Tschann, JM; Wara, DW; Boyce, WT. Life events, frontal electroencephalogram laterality, and functional immune status after acute psychological stressors in adolescents. *Psychosom. Med.* 1997; 59, 178-86.

Lindemann, RA; Park, NH. Inhibition of human lymphokine-activated killer activity by smokeless tobacco (snuff) extract. *Arch Oral Biol.* 1988; 33, 317-21.

Lowin, B; Hahne, M; Mattmann, C; Tschopp, J. Cytolytic T-cell cytotoxicity is mediated through perforin and Fas lytic pathways. *Nature* 1994; 370, 650-652.

Matsumoto, Y; Mishima, K; Satoh, K; Tozawa, T; Mishima, Y; Shimizu, T; Hishikawa, Y. Total sleep deprivation induces an acute and transient increase in NK cell activity in healthy young volunteers. *Sleep* 2001; 24, 804-809.

Meliska, CJ; Stunkard, ME; Gilbert, DG; Jensen, RA; Martinko, JM. Immune function in cigarette smokers who quit smoking for 31 days. *J Allergy Clin Immunol.* 1995; 95, 901-10.

Miles, MP; Mackinnon, LT; Grove, DS; Williams, NI.; Bush, JA; Marx, JO; Kraemer, WJ; Mastro, AM. The relationship of natural killer cell counts, perforin mRNA and CD2 expression to post-exercise natural killer cell activity in humans. *Acta Physiol. Scand.* 2002; 174, 317-325.

Moldofsky, H; Lue, FA; Davidson, JR; Gorczynski, R. Effects of sleep deprivation on human immune functions. *FASEB J.* 1989; 3, 1972-1977.

Moleman, P; Tulen, JH; Blankestijn, PJ; Man in 't Veld, AJ; Boomsma, F. Urinary excretion of catecholamines and their metabolites in relation to circulating catecholamines. Six-hour infusion of epinephrine and norepinephrine in healthy volunteers. *Arch Gen Psychiatry.* 1992; 49, 568-72.

Mori, H; Nishijo, K; Kawamura, H; Abo, T. Unique immunomodulation by electro-acupuncture in humans possibly via stimulation of the autonomic nervous system. *Neurosci. Lett.* 2002; 320, 21.

Morimoto, K. Life-style and genetic factors that determine the susceptibility to production of chromosome damage. In: G. Oge and A.T. Natarajan, Editors, *Chromosome aberration: basic and applied aspects.* Berlin: Springer-Verlag, 1990; pp. 287–301.

Morimoto, K; Takeshita, T; Inoue-Sakurai, C; Maruyama, S. Lifestyles and mental health status are associated with natural killer cell and lymphokine-activated killer cell activities. *Sci. Total Environ.* 2001; 270, 3-11.

Morimoto, K; Takeshita, T; Nanno, M; Tokudome, S; Nakayama, K. Modulation of natural killer cell activity by supplementation of fermented milk containing Lactobacillus casei in habitual smokers. *Prev Med.* 2005; 40, 589-94.

Morita, E; Fukuda, S; Nagano, J; Hamajima, N; Yamamoto, H; Iwai, Y; Nakashima, T; Ohira, H; Shirakawa, T. Psychological effects of forest environments on healthy adults: Shinrin-yoku (forest-air bathing, walking) as a possible method of stress reduction. *Public Health* 2007; 121, 54-63.

Motivala, SJ; Dang, J; Obradovic, T; Meadows, GG; Butch, AW; Irwin MR. Leptin and cellular and innate immunity in abstinent alcoholics and controls. *Alcohol Clin Exp Res.* 2003; 27, 1819-24.

Nagel, JE; Collins, GD; Adler, WH. Spontaneous or natural killer cytotoxicity of K562 erythroleukemic cells in normal patients. *Cancer Res.* 1981; 41, 2284-8.

Nair, MP; Kronfol, ZA; Schwartz, SA. Effects of alcohol and nicotine on cytotoxic functions of human lymphocytes. *Clin Immunol Immunopathol.* 1990; 54; 395-409.

Nakachi, K; Imai, K. Environmental and physiological influences on human natural killer cell activity in relation to good health practices. *Jpn J Cancer Res.* 1992; 83, 798-805.

Nakamura, H; Nagase, H; Yoshida, M; Ogino, K. Natural killer (NK) cell activity and NK cell subsets in workers with a tendency of burnout. *J Psychosom Res.* 1999; 46, 569-78.

Nakamura, H; Matsuzaki, I; Sasahara, S; Hatta, K; Nagase, H; Oshita, Y; Ogawa, Y; Nobukuni, Y; Kambayashi, Y; Ogino, K. Enhancement of a sense of coherence and natural killer cell activity which occurred in subjects who improved their exercise habits through health education in the workplace. *J Occup Health.* 2003; 45, 278-85.

Nieman, DC. Special feature for the Olympics: effects of exercise on the immune system: exercise effects on systemic immunity. *Immunol. Cell Biol.* 2000; 78, 496-501.

Newman, LS; Kreiss, K; Campbell, PA. Natural killer cell tumoricidal activity in cigarette smokers and in silicotics. *Clin Immunol Immunopathol.* 1991; 60, 399-411.

Ochshorn-Adelson, M; Bodner, G; Toraker, P; Albeck, H; Ho, A; Kreek, MJ. Effects of ethanol on human natural killer cell activity: in vitro and acute, low-dose in vivo studies. *Alcohol Clin. Exp. Res.* 1994; 18, 1361-1367.

Ogawa, K; Takamori, Y; Suzuki, K; Nagasawa, M; Takano, S; Kasahara, Y; Nakamura, Y; Kondo, S; Sugamura, K; Nakamura, M; Nagata, K. Granulysin in human serum as a marker of cell-mediated immunity. *Eur J Immunol* 2003; 33, 1925-1933.

Ohtsuka, Y; Yabunaka, N; Takayama, S. Shinrin-yoku (forest-air bathing and walking) effectively decreases blood glucose levels in diabetic patients. *Int. J. Biometeorol.* 1998; 41, 125-127.

Okada, S; Li, Q; Whitin, JC; Clayberger, C; Krensky, AM. Intracellular mediators of granulysin-induced cell death. *J. Immunol.* 2003; 171, 2556-2562.

Park BJ, Tsunetsugu Y, Kasetani T, Hirano H, Kagawa T, Sato M, Miyazaki Y. Physiological effects of Shinrin-yoku (taking in the atmosphere of the forest)--using salivary cortisol and cerebral activity as indicators. J Physiol Anthropol. 2007, 26, 123-8.

Pedersen, BK; Tvede, N; Christensen, LD; Klarlund, K; Kragbak, S; Halkjr-Kristensen, J. Natural killer cell activity in peripheral blood of highly trained and untrained persons. *Int J Sports Med.* 1989; 10, 129-31.

Perney, P; Portales, P; Corbeau, P; Roques, V; Blanc, F; Clot, J. Specific alteration of peripheral cytotoxic cell perforin expression in alcoholic patients: a possible role in alcohol-related diseases. *Alcohol Clin. Exp. Res.* 2003; 27, 1825-30.

Phillips, B; Marshall, ME; Brown, S; Thompson, JS. Effect of smoking on human natural killer cell activity. *Cancer* 1985; 56, 2789-92.

Redwine, L; Dang, J; Hall, M; Irwin, M. Disordered sleep, nocturnal cytokines, and immunity in alcoholics. *Psychosom Med.* 2003; 65, 75-85.

Rivoltini, L; Cattoretti, G; Arienti, F; Mastroianni, A; Melani, C; Colombo, MP; Parmiani, G. The high lysability by LAK cells of colon-carcinoma cells resistant to doxorubicin is associated with a high expression of ICAM-1, LFA-3, NCA and a less-differentiated phenotype. *Int J Cancer.* 1991; 47, 746–754.

Rohde, T; MacLean, DA; Pedersen, BK. Effect of glutamine supplementation on changes in the immune system induced by repeated exercise. *Med Sci Sports Exerc.* 1998; 30, 856-62.

Sahiar, F; Mohler, SR. Economy class syndrome. *Aviat Space Environ Med.* 1994; 65, 957-60.

Sakami, S; Ishikawa, T; Kawakami, N; Haratani, T; Fukui, A; Kobayashi, F; Fujita, O; Araki, S; Kawamura, N. Coemergence of insomnia and a shift in the Th1/Th2 balance toward Th2 dominance. *Neuroimmunomodulation.* 2002-2003; 10, 337-43.

Savas, B; Arslan, G; Gelen, T; Karpuzoglu, G; Ozkaynak, C. Multidrug resistant malignant melanoma with intracranial metastasis responding to immunotherapy. *Anticancer* Res. *1999; 19, 4413–4420.*

Savas, B; Kerr, PE; Ustun, H; Cole, SP; Pross, HF. Lymphokine-activated killer cell susceptibility and multidrug resistance in small cell lung carcinoma. *Anticancer Res.* 1998; 18, 4355–4361.

Savas, B; Cole, SP; Tsuruo, T; Pross, HF. P-glycoprotein-mediated multidrug resistance and lymphokine-activated killer cell susceptibility in ovarian carcinoma. *J Clin Immunol.* 1996; 16, 348–357.

Schaberg, T; Theilacker, C; Nitschke, OT; Lode, H. Lymphocyte subsets in peripheral blood and smoking habits. *Lung* 1997; 175, 387-394.

Shaw, HM; Milton, GW; McCarthy, WH; Farago, GA; Dilworth, P. Effect of smoking on the recurrence of malignant melanoma. *Med J Aust.* 1979; 1, 208-9.

Shinkai,Y; Takio, K; Okumura, K. Homology of perforin to the ninth component of complement (C9). *Nature* 1988; 334, 525-527.

Simon, MM; Hausmann, M; Tran, T; Ebnet, K; Tschopp, J; ThaHla, R; Mullbacher, A. In vitro- and ex vivo-derived cytolytic leukocytes from granzyme A x B double knockout mice are defective in granule-mediated apoptosis but not lysis of target cells. *J Exp Med.* 1997; 186, 1781-6.

Sluiter, JK; van der Beek, AJ; Frings-Dresen, MH. Work stress and recovery measured by urinary catecholamines and cortisol excretion in long distance coach drivers. *Occup Environ Med.* 1998; 55, 407-13.

Smyth, MJ; Kelly, JM; Sutton, VR; Davis, JE; Browne KA; Sayers, TJ, Trapani, JA. Unlocking the secrets of cytotoxic granule proteins. *J. Leukoc. Biol.* 2001; 70, 18-29.

Staats, R;, Balkow, S; Sorichter, S; Northoff, H; Matthys, H;, Luttmann, W; Berg, A; Virchow, JC. Change in perforin-positive peripheral blood lymphocyte (PBL) subpopulations following exercise. *Clin. Exp. Immunol.* 2000; 120, 434-439.

Stenger, S; Hanson, DA; Teitelbaum, R; Dewan, P; Niazi, KR; Froelich, CJ; Ganz, T; Thoma-Uszynski, S; Melian, A; Bogdan, C; Porcelli, SA; Bloom BR; Krensky, AM.; Modlin, RL. An antimicrobial activity of cytolytic T cells mediated by granulysin. *Science* 1998; 282, 121-125.

Suzui, M; Kawai, T; Kimura, H; Takeda, K; Yagita, H; Okumura, K; Shek, PN; Shephard, RJ. Natural killer cell lytic activity and CD56 (dim) and CD56 (bright) cell distributions during and after intensive training. *J Appl Physiol.* 2004; 96, 2167-73.

Takeuchi, M; Nagai, S; Izumi, T. Effect of smoking on natural killer cell activity in the lung. *Chest.* 1988; 94, 688-93.

Takeuchi, M; Nagai, S; Nakajima, A; Shinya, M; Tsukano, C; Asada, H; Yoshikawa, K; Yoshimura, M; Izumi, T. Inhibition of lung natural killer cell activity by smoking: the role of alveolar macrophages. *Respiration.* 2001; 68, 262-7.

Toda, M; Makino, H; Kobayashi, H; Nagasawa, S; Kitamura, K; Morimoto, K. Medical assessment of the health effects of short leisure trips. *Arch Environ Health* 2004; 59, 717-24.

Tanigawa, T; Araki, S; Morimoto, K; Yokoyama, K. Effects of physical exercise on natural killer cell activity in healthy men in relation to life-style. In: Araki, S. (Ed.), *Behavioral*

Medicine: An Integrated Biobehavioral Approach to Health and Illness. Amsterdam/New York: Elsevier, 1992; pp. 79-84.

Tollerud, DJ; Clark, JW; Brown, LM; Neuland, CY; Mann, DL; Pankiw-Trost, LK; Blattner, WA; Hoover, RN. Association of cigarette smoking with decreased numbers of circulating natural killer cells. *Am. Rev. Respir. Dis.* 1989a; 139, 194-198.

Tollerud, DJ; Clark, JW; Brown, LM; Neuland, CY; Mann, DL; Pankiw-Trost, LK; Blattner, WA; Hoover, RN. The effects of cigarette smoking on T cell subsets. A population-based survey of healthy caucasians. *Am. Rev. Respir. Dis.* 1989b; 139, 1446-1451.

Trinchieri, G. Biology of natural killer cells. *Adv Immunol.* 1989; 47, 187-376.

Tuschl, H; Weber, E; Kovac, R. Investigations on immune parameters in welders. *J Appl Toxicol.* 1997; 17, 377-83.

Tvede, N; Kappel, M; Halkjaer-Kristensen, J; Galbo, H; Pedersen, BK. The effect of light, moderate and severe bicycle exercise on lymphocyte subsets, natural and lymphokine activated killer cells, lymphocyte proliferative response and interleukin 2 production. *Int J Sports Med.* 1993; 14, 275-82.

Ullum, H; Palmo, J; Halkjaer-Kristensen, J; Diamant, M; Klokker, M; Kruuse, A; LaPerriere, A; Pedersen, BK. The effect of acute exercise on lymphocyte subsets, natural killer cells, proliferative responses, and cytokines in HIV-seropositive persons. *J Acquir Immune Defic Syndr.* 1994; 7, 1122-33.

van der Beek, AJ; Meijman, TF; Frings-Dresen, MH; Kuiper, JI; Kuiper, S. Lorry drivers' work stress evaluated by catecholamines excreted in urine. *Occup Environ Med.* 1995; 52, 464-9.

Van Kaer, L. NKT cells: T lymphocytes with innate effector functions. *Curr Opin Immunol.* 2007; 19(3), 354-64.

Venkatraman, JT; Pendergast, DR. Effect of dietary intake on immune function in athletes. *Sports Med.* 2002; 32(5), 323-37.

Watzl, B; Bub, A; Briviba, K; Rechkemmer, G. Acute intake of moderate amounts of red wine or alcohol has no effect on the immune system of healthy men. *Eur J Nutr.* 2002; 41, 264-70.

Watzl, B; Bub, A; Pretzer, G; Roser, S; Barth, SW; Rechkemmer, G. Daily moderate amounts of red wine or alcohol have no effect on the immune system of healthy men. *Eur J Clin Nutr.* 2004; 58, 40-5.

Waterhouse, J; Reilly, T; Edwards, B. The stress of travel. *J Sports Sci.* 2004; 22, 946-66.

Yamaguchi, M; Deguchi, M; Miyazaki, Y. The effects of exercise in forest and urban environments on sympathetic nervous activity of normal young adults. *J. Int. Med. Res.* 2006; 34, 152-9.

Yovel, G; Sirota, P; Mazeh, D; Shakhar, G; Rosenne, E; Ben-Eliyahu, S. Higher natural killer cell activity in schizophrenic patients: the impact of serum factors, medication, and smoking. *Brain Behav Immun.* 2000; 14, 153-69.

Zeidel, A; Beilin, B; Yardeni I; Mayburd E; Smirnov G; Bessler H. Immune response in asymptomatic smokers. *Acta Anaesthesiol Scand.* 2002; 46, 959-64.

Zhang, D; Beresford, PJ; Greenberg, AH.; Lieberman, J. Granzymes A and B directly cleave lamins and disrupt the nuclear lamina during granule-mediated cytolysis. *Proc. Natl. Acad. Sci. USA* 2001a; 98, 5746-5751.

Zhang, D; Pasternack, MS; Beresford, PJ; Wagner, L; Greenberg, AH; Lieberman, J. Induction of rapid histone degradation by the cytotoxic T lymphocyte protease Granzyme A. *J. Biol. Chem.* 2001b; 276, 3683-3690.

In: Natural Killer T-Cells: Roles, Interactions and Interventions   ISBN: 978-1-60456-287-3
Editor: Nathan V. Fournier, pp. 81-102   © 2008 Nova Science Publishers, Inc.

Chapter 3

Organophosphorus Compounds Inhibit Natural Killer Cell Activity

Qing Li[*]
Department of Hygiene and Public Health, Nippon Medical School
1-1-5 Sendagi, Bunkyo-ku, Tokyo 113-8602, Japan

Abstract

Organophosphorus pesticides (OPs) are widely used throughout the world as insecticides in agriculture and eradicating agents for termites around homes. The main toxicity of OPs is neurotoxicity, which is caused by the inhibition of acetylcholinesterase. OPs also affect immune response including effects on antibody production, IL-2 production, T cell proliferation, Th1/Th2 cytokine profiles, decrease of CD5 cells, and increases of CD26 cells and autoantibodies, inhibitions of lymphokine-activated killer and cytotoxic T lymphocytes activities. However, there have been few papers investigating the mechanism of OP-induced inhibition of natural killer (NK) activity. This study reviews the effect of organophosphorus compounds on NK activity and the mechanism of organophosphorus compound-induced inhibition of NK activity. It has been reported that NK cells induce cell death in tumor or virus-infected target cells by two main mechanisms. The first mechanism is direct release of cytolytic granules that contain the pore-forming protein perforin, several serine proteases termed granzymes, and granulysin by exocytosis to kill target cells, which is called the granule exocytosis pathway. The second mechanism is mediated by the Fas ligand (Fas-L)/Fas pathway, in which FasL (CD95L), a surface membrane ligand of the killer cell cross links with the target cell's surface death receptor Fas (CD95) to induce apoptosis of the target cells. We have previously found that organophosphorus compounds including OPs significantly inhibit human and animal NK activity both in vitro and in vivo. Moreover, we have found that OPs inhibit NK activity by at least the following three mechanisms: 1) OPs impair the granule exocytosis pathway of NK cells by inhibiting the activity of granzymes, and by decreasing the intracellular level of perforin, granzymes and granulysin, which was mediated by inducing degranulation of NK cells and by inhibiting the transcript of

---
[*] E-mail address: qing-li@nms.ac.jp, Tel: +81-3-3822-2131, Fax: +81-3-5685-3065.

mRNA of perforin, granzyme A and granulysin. 2) OPs impair the FasL/Fas pathway of NK cells, as investigated by using perforin-knockout mice, in which the granule exocytosis pathway of NK cells does not function and only the FasL/Fas pathway remains functional. 3) OPs induce apoptosis of NK cells, which is at least partially mediated by activation of intracellular caspase-3.

Key words: apoptosis, granulysin, granzyme, NK, organophosphorus pesticide, perforin.

Introduction

Organophosphorus pesticides (OPs) are one of the main classes of insecticides, in use since the mid 1940s. OPs are widely used throughout the world as insecticides in agriculture and as agents for eradicating termites around homes [1-4]. There is still a large of quantity of OPs on the market in Japan [5]. OPs can exert significant adverse effects in non-target species including humans, and are potent inhibitors of serine esterases, such as acetylcholinesterase and serum cholinesterase [1-4, 6]. The main toxicity of OPs is neurotoxicity, which is caused by the inhibition of acetylcholinesterase [1-4, 6, 7]. It has been reported that OPs affect immune response including effects on neutrophil function [8], macrophage [9-12], antibody production [13, 14], interleukin (IL)-2 production [15], serum complement [16], and T-cell proliferation induced by IL-2 [17], concanavalin A, and phytohemagglutinin [18] in animals and humans. Thrasher et al. [19, 20] have reported that higher-than-usual frequencies of allergies and sensitivities to antibiotics, a decrease in CD5 cells, and increases in CD26 cells and autoantibodies were found in patients following chlorpyrifos exposure. Increased expression of CD26 cells and decreased expression of CD5 cells are associated with autoimmunity, where an individual's immune system acts against itself, rather than against infections [21]. Rodgers has also reported that oral administration of malathion increases levels of anti-dsDNA antibodies in MRL-lpr mice [22]. Exposure to chlorpyrifos is associated with multiple chemical sensitivity [23, 24].

Potential Mechanism of OP-Induced Immunotoxicity

There are very few papers dealing with the mechanism of OP-induced immunotoxicity. Gallowway and Habdy [25] have reviewed the potential mechanism of OP-induced immunotoxicity, including the direct and indirect effects. OPs may inhibit any of the serine hydrolase class of enzyme in the immune system, including the complement [12] and thrombin systems, which influence immune function. The OP-induced inhibition of neuropathy target esterase in lymphocytes may lead to structural or functional changes in lymphocytes [26]. Histopathological damage to lymphoid tissues resulting from phosphorylation, oxidative damage, or altered neural function induced by OPs could hinder the development and viability of lymphocytes [27]. On the other hand, Osicka-Koprowska et al. [28] have reported that chlorfenvinphos induces a significant increase in corticosterone in rat plasma, suggesting that indirect immune alterations may be mediated by cholinergic responses or by stress following neurotoxic doses. Videira et al. [29] have reported that malathion, methylparathion, and parathion can directly damage cell membranes by affecting membrane lipid physicochemical properties, which may indirectly influence the function of immune cells. We have also found that organophosphorus compounds induce an increase in sister chromatid exchanges (SCEs) in human lymphocytes, which could indirectly influence the function of lymphocytes [30, 31].

Effect of OPs on Th1/Th2 Cytokine Profiles

Exposure to OP has been linked to asthma or asthma-related symptoms in a small number of epidemiologic studies. Hoppin et al. [32] have reported that OP exposure in the preceding year is related to wheezing in a study of pesticide applicators in Iowa and North Carolina. Respiratory asthma-like symptoms were associated with exposures to OPs in occupational and environmental settings among villagers in rural China [33]. Asthma is characterized by chronic inflammation in the airways and a predominance of CD4+ T-helper 2 (Th2) cells that secrete IL-4, IL-5, and IL-13 cytokines [34]. Th2 cells contribute to the immunopathogenesis of asthma by recruiting eosinophils and mast cells to the airways [34-36] and by inducing B-cells to produce immunoglobulin E antibodies [37]. Conversely, T-helper 1 (Th1) cells that secrete interferon (IFN)-γ are thought to protect against the development of asthma by regulating Th2 cytokine production, although a mixed Th1/Th2 pattern has been reported [38]. Patients with allergies and asthma are more likely to have elevated levels of the Th2 cytokines IL-4 and IL-5 and reduced levels of the Th1 cytokines IFN-γ and tumor necrosis factor (TNF)-β [34, 39, 40]. These findings suggest that OPs may have effects on the Th1/Th2 cytokine profiles. Thus, Duramad et al [41] investigated the effect of chlorpyrifos on the expression of Th1/Th2 cytokines in human blood by in vitro treatment and found that although chlorpyrifos did not induce cytokine expression in vitro, it did increase lipopolysaccharide (LPS)-dependent induction of IFN-γ. Moreover, Duramad et al [42] have found that early environmental exposure to OPs may affect intracellular Th1/Th2 cytokine profiles in 24-month-old children living in an agricultural area of Salinas Valley, California.

Effect of Organophosphorus Compounds on Cytolytic Killer Cells

On March 20, 1995, the nerve gas sarin (isopropylmethylphosphonofluoridate) was used in a terrorist attack by the members of the Aum Shinrikyo cult on subway trains in the Tokyo metropolitan area, killing 12 people and injuring more than 5000 [43]. Sarin is an OP nerve agent and a strong cholinesterase inhibitor [44]. To monitor the genetic aftereffects of sarin exposure, we measured SCEs in peripheral blood lymphocytes of the victims and found that the frequency of SCEs in lymphocytes was significantly higher in the victims than in the control group [30, 45]. We also found that the by-products generated during sarin synthesis, i.e., diisopropyl methylphosphonate (DIMP), diethyl methylphosphonate (DEMP), and ethyl isopropyl methylphosphonate (EIMP), also induce SCEs in human lymphocyte *in vitro* [30]. On the other hand, in the Tokyo sarin attack, the victims were also exposed to *N,N*-diethylaniline, a stabilizing reagent of sarin synthesis. We first found that *N,N*-diethylaniline significantly increase the frequency of SCEs of human lymphocytes *in vitro* [46]. Because DIMP, DEMP and *N,N*-diethylaniline induce a high frequency of SCEs of lymphocytes, we have speculated that DIMP, DEMP, and *N,N*-diethylaniline also affect the function of lymphocytes. Thus, we investigated the effects of DIMP, DEMP, and *N,N*-diethylaniline on natural killer cell (NK) and cytotoxic T lymphocyte (CTL) activities both *in vitro* and *in vivo* and found that DIMP, DEMP, and *N,N*-diethylaniline significantly inhibit human and murine NK cells and murine CTL activities both *in vitro* and *in vivo*, suggesting a relationship between the increased SCEs in lymphocytes and decreased NK cell, CTL, and lymphokine-activated killer (LAK) cell activities induced by DIMP, DEMP, and *N,N*-diethylaniline [31,47]. Both DIMP and DEMP are potent inhibitors of serine esterases, such as acetylcholinesterase and serum cholinesterase, which are similar to OPs in toxicity [48]. Thus, we speculate that OPs also may inhibit NK cell and CTL activities as do DIMP and DEMP. To clarify whether OPs also affect NK cell and CTL activities, we first investigated five OPs — dimethyl 2,2-dichlorovinyl phosphate (DDVP, dichlorvos), dimethyl 2,2,2-trichloro-hydroxyethylphosphonate (DEP), dimethoate (DMTA), acephate and S-2-ethylsulfinyl-1-methylethyl O,O-dimethyl phosphorothioate (ESP) — on human NK cell activity. We found that all five OPs significantly decreased human NK cell activity in a dose-dependent manner and that the strength of inhibition differed among the five OPs in the following order: DDVP > DMTA ≥ DEP ≥ ESP > acephate [49] (Fig. 1). Then we investigated the effect of DDVP on murine splenic NK cell, LAK cell, and CTL activities and human NK and LAK activities. DDVP significantly decreased human NK cell [49-52], LAK [49], and murine NK, LAK and CTL activities *in vitro* [49] and *in vivo* [53] in a dose-dependent manner, and the degree of decrease in these activities differed among the effector cells investigated. The order was as follows: human NK cells > murine NK cells > murine CTLs > murine LAK cells > human LAK cells [49].

Table 1. Summary of the effect of organophosphorus compounds on cytolytic killer cells (58, 59)

Targets (Cells)	Parameters	Effects	Human/animal	References
NK cells	NK activity	↓↓↓↓↓↓↓↓↓↓↓	Human/mice	31, 49-57, 80
	Granzymes activity	↓↓↓	Human	49, 73, 74
	Expressions of perforin, granulysin, GrA, GrB, Gr3/K	↓↓↓	Human	50-52
	FasL/Fas pathway	↓	Mice	53
	Apoptosis	↑ (Positive)	Human	80
LAK cells	LAK activity	↓↓	Human/mice	49, 53
	FasL/Fas pathway	↓	Mice	53
T cells	CTL activity	↓↓↓↓↓	Human/mice	31, 49, 53, 55-57
	FasL/Fas pathway	↓	Mice	53
	Apoptosis	↑↑ (Positive)	Mice	81, 82

↓: Inhibition/decrease; ↑: Increase/Activation (Induction). Numbers of the arrow show the number of references.

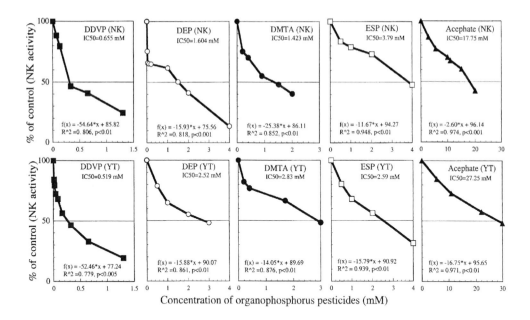

Figure 1. Effect of OPs on human NK cell activity and YT cell activity *in vitro*. IC50: inhibitory concentration of 50% NK cell activity. The YT cell is a human NK cell line. Cited from Li et al., Toxicology 2002, 172, 181-190.

Zabrodskii and Germanchuk [54] have also reported that subcutaneous injection of DDVP at 0.2 of the median lethal dose (LD50) and 0.8 of the LD50 significantly inhibited NK cell activity and antibody-dependent cell cytotoxicity in Wistar rats. Rodgers et al. have reported that O,O,S-trimethyl phosphorothioate, an impurity in technical formulations of malathion, inhibited human NK cell activity *in vitro* [55] and murine and human CTL activity *in vivo* and *in vitro* [56,57]. Table 1 summarizes the effect of organophosphorus compounds on cytolytic killer cells in humans and animals [58, 59].

However, there have been few studies of the mechanisms of OP-induced inhibition of NK cell, LAK cell, and CTL activities. We review the new mechanisms of OP-induced inhibition of cytolytic activity of killer cells in the present study.

Mechanisms of OP-Induced Inhibition of NK Activity

1. OPs Impair the Granule Exocytosis Pathway of NK Cells

It has been reported that NK cells induce tumor or virus-infected target cell death by two main mechanisms [58-62]. The first mechanism is the direct release of cytolytic granules that contain the pore-forming protein perforin, several serine proteases termed granzymes [63], and granulysin [64] by exocytosis to kill target cells. The second mechanism is mediated by the Fas ligand (FasL)/Fas pathway, in which FasL (CD95L), a surface membrane ligand of the killer cell cross links with the target cell's surface death receptor Fas (CD95) to induce apoptosis of the target cells [53, 60-62, 65]. Human NK cells, LAK cells and CTLs have been shown to express five granzymes. Granzyme A (GrA) is expressed in NK cells, phytohemagglutinin- or CD3-stimulated T cells, and gamma/delta T cells and has a trypsin-like specificity, which cleaves on the carboxyl side of basic residues, such as arginine and lysine [66, 67]. GrB is expressed in NK cells, PHA-or CD3-stimulated T cells, gamma/delta T cells, and cleaves on the carboxyl side of aspartic acid residues [67, 68]. Gr3/K is expressed in T cells and IL-2 or ConA-stimulated T cells, NK cells and PBL and is trypsin-like (cleavage after basic residues) [69, 70]. GrH is expressed in IL-2- or PHA-stimulated PBLs and CTLs, and prefers cleavage after hydrophobic residues, such as phenylalanine [71]. GrM is expressed in NK cells and gamma/delta T cells, and cleaves on the carboxyl side of methionine, leucine, or norleucine [67, 72].

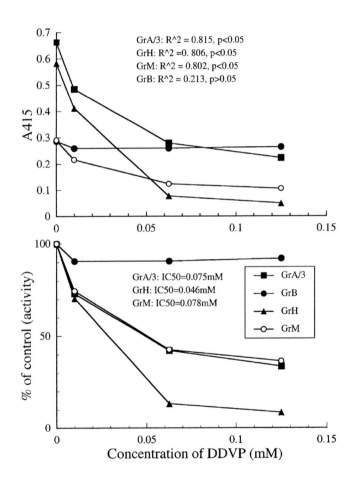

Figure 2. Inhibitory effect of DDVP on activity of human granzymes A, B, 3, H, M. IC50: inhibitory concentration of 50% activity of granzymes. Cited from Li et al., Toxicology 2002, 172, 181-190.

OPs are potent inhibitors of serine esterases, such as acetylcholinesterase and serum cholinesterase [1-4, 6], and granzymes are also serine esterases (proteases) [63, 66-72]. Thus, we speculate that the decrease in NK cell, LAK cell, and CTL activities by OPs may be mediated by the inhibition of serine proteases (granzymes), which are released from NK cell and CTL granules by exocytosis when target cells conjugate with the effector cells. To explore the underlying mechanism of the decrease in cytolytic activity in killer cells, we investigated the effects of DDVP on the enzymatic activity of human granzymes and found that DDVP significantly inhibits the enzymatic activity of human GrA, Gr3, GrH, and GrM in a dose-dependent manner. The IC50 (inhibitory concentration of 50% granzyme activity) values were 0.05 mM for GrA and Gr3, 0.03 mM for GrH and 0.05 mM for GrM (Fig. 2). To support our hypothesis that OP-induced inhibition of the cytolytic activity of killer cells is mediated by the inhibition of granzymes (serine proteases), we investigated the effect of 4-(2-aminoethyl) benzenesulfonyl fluoride-HCl (*p*-ABSF), an inhibitor of serine proteases, on NK cell, LAK cell, and CTL activities. We found that *p*-ABSF significantly decreased the activities of human and murine NK cells and LAK cells and of a murine CTL line in a dose-dependent manner, and that the degree of decrease in those activities also differed among the effector cells in the following order: human NK > murine NK > murine CTL line > murine LAK > human LAK. This order coincides with the results obtained with DDVP, suggesting that DDVP and *p*-ABSF have a common inhibiting mechanism on NK cell, LAK cell and CTL activities. In addition, the decreases in NK cell, LAK cell, and CTL activities induced by *p*-ABSF+DDVP were greater than those by either *p*-ABSF alone or DDVP alone in the same concentration, suggesting that DDVP and *p*-ABSF have an additive inhibitory effect on NK cell, CTL, and LAK cell activities. Mahrus and Craik [73] also reported that diphenyl phosphonates such as phosphonates Bio-x-IGN (AmPhg)^P-(OPh)$_2$ and Bio-x-IEPD^P-(OPh)$_2$ significantly inhibit human GrA and GrB, respectively. Oragnophosphorus compound, DFP (diisopropyl phosphofluoridate), which is a general serine protease inhibitor was observed to block the cytolytic activity of cytotoxic lymphocytes [74]. Taken together, the above-mentioned findings indicate that OPs significantly decrease NK cell, LAK cell, and CTL activities *in vitro,* at least partially mediated by granzyme inhibition [49].

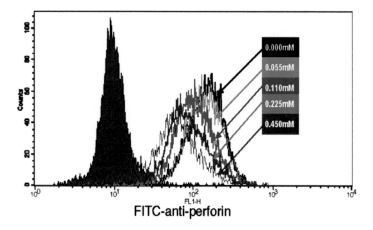

FITC-anti-perforin

Figure 3. Effect of DDVP on the expression of perforin in human NK-92CI after 15-hour in vitro treatment. The X axis shows the fluorescent intensity of fluorescein isothiocyanate (FITC)-anti-perforin, which represents the intracellular level of perforin, and the Y axis shows the counts of NK cells. The solid histogram shows the control stained with FITC-mouse IgG2b (isotype control), and the blank histograms show the results of staining with FITC-mouse anti-human perforin after treatment with DDVP at 0 (black), 0.055 (green), 0.110 (blue), 0.225 (orange) and 0.452 (red) mM from the right to the left, respectively. Cited from Li et al., Toxicology 2005, 213, 107-116.

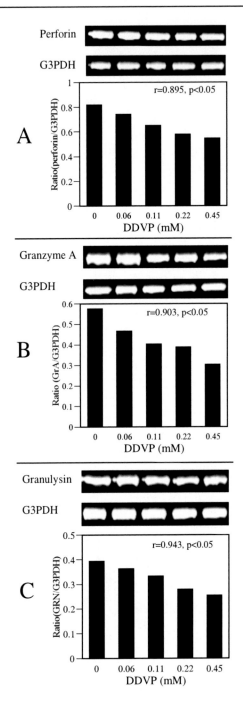

Figure 4. Effect of DDVP on the transcription of mRNA of perforin (A), granzyme A (B) and granulysin (C) in NK-92CI after 15 h in vitro treatment. Photos obtained from the PCR products with 30 cycles which were applied on a 7.5% polyacrylamide gel with a 5% stacking gel and the PCR products were quantified by Fluor Imager 595. All the results are indicated by the fluorescent densitometric ratio of mRNA of each product of perforin, granzyme A and granulysin to that of G3PDH. The applied volumes for electrophoresis are 5ul per lane for perforin, granzyme A and granulysin and 2ul per lane for G3PDH. Experiments were repeated at least 3 times with very similar results. Cited from Li et al., Toxicology 2005, 213, 107-116.

To investigate whether OPs also affect the expression of granzyme, granulysin, and perforin, we treated a human NK cell line, NK-92, with DDVP *in vitro* and then analyzed the expressions of granzyme, granulysin, and perforin with flow cytometry and reverse transcriptase polymerase chain reaction. We found that DDVP significantly decreased the expression of perforin (Fig. 3), granulysin, and granzyme A, B, 3/K in NK-92CI and NK-92 MI cells in a dose-dependent manner [50-52]. DDVP also has a modest but significant inhibitory effect on the transcription of the mRNAs of perforin, granzyme A, and granulysin (Fig. 4). Moreover, we found that the decreases in perforin, granzyme A, and granulysin in the granules of NK-92CI cells parallel a similar pattern determined by immunocytochemical analysis, which strongly suggests the possibility of degranulation [50].

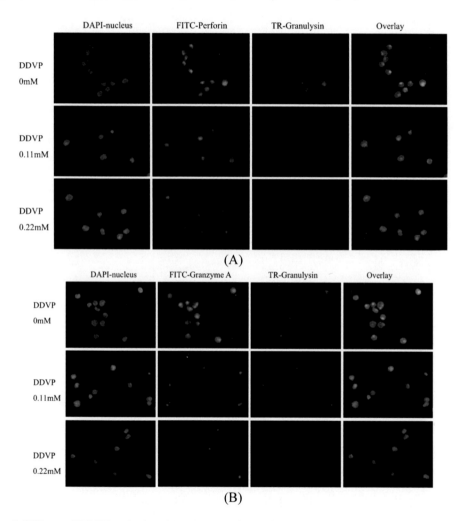

Figure 5. Effects of DDVP at 0.11 and 0.22 mM on intracellular perforin/granulysin (A) and granzyme A/granulysin (B) in NK-92CI cells after 15-hour in vitro treatment. The NK-92CI cells were fixed/permeablized with Cytofix/cytoperm solution, after which double-staining of perforin/granulysin and granzyme A/granulysin was performed. The intracellular perforin and granzyme A were stained with FITC-anti-human perforin and granzyme A, respectively. Intracellular granulysin was first stained with rabbit anti-human granulusin polyclonal antibody, then stained with TR-goat anti-rabbit IgG. Cited from Li et al., Toxicology 2005, 213, 107-116.

Taken together these findings indicate that DDVP inhibits the enzymatic activity of granzymes [49], the expression of granzymes, granulysin, and perforin in human NK cells, and the induction of degranulation of NK cells [50,51] (Fig. 5).

2. OPs Impair the FasL/Fas Pathway of NK Cells

Only one paper has investigated whether OPs affect the FasL/Fas pathway of killer cells using perforin-knockout (PKO) mice [53]. It has been reported that the granule exocytosis pathway in PKO mice does not function against Fas antigen-negative target cells [53,65,75] and that the NK cells, CTLs, and LAK cells of PKO mice kill targets only by the FasL/Fas pathway [53,65,75]. Thus, we used PKO mice to investigate the effect of DDVP on the FasL/Fas pathway by determining the NK cell, CTL, and LAK cell activities in PKO mice.

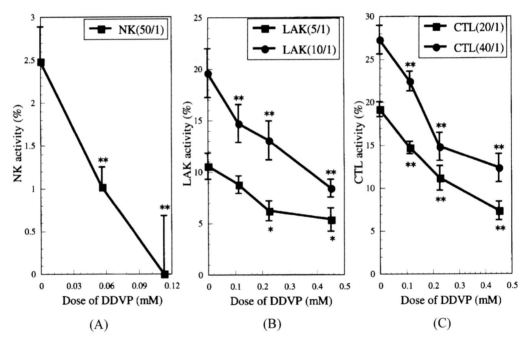

Figure 6. Effect of DDVP on NK cell (A), LAK cell (B), and CTL (C) activities of PKO mice *in vitro*. Cited from Li et al., Toxicology 2004, 204, 41-50.

In this study, we found that DDVP significantly decreased the NK cell, CTL, and LAK cell activities of PKO mice in a dose-dependent manner (Fig. 6) and that the CTL and LAK cell activities of PKO mice were significantly blocked by an anti-FasL antibody, suggesting that DDVP and the anti-FasL antibody have the same or a similar mechanism of inhibiting LAK cell and CTL activities. Moreover, DDVP decreases the expression of Fas antigen on YAC-1 cells (a target cell in NK cell activity assay) and the expression of FasL on LAK cells in a dose-dependent manner (Fig. 7). Taken together, these findings indicate that the DDVP-induced inhibition of NK cell, LAK cell, and CTL activities in PKO mice is mediated by the impairment of the FasL/Fas pathway [53].

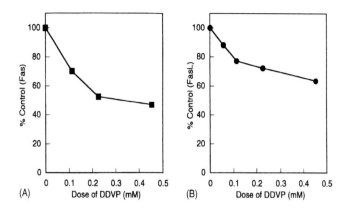

Figure 7. Effect of DDVP on the expression of Fas antigen on the surface of YAC-1 cells (A) and on the expression of FasL on the surface of LAK cells (B). Cited from Li et al., Toxicology 2004, 204, 41-50.

3. OPs Induce Apoptosis of NK Cells

It has been reported that OPs induced apoptosis in rat primary cortical neurons [76], in SH-SY5Y human neuroblastoma cells [77], and in murine preimplantation embryos [78]. To explore the mechanism of OP-induced inhibition of NK activity, we investigated whether OPs induced apoptosis in human immune cells and examined the underlying mechanism [79]. We first treated human immune cells, a human monocyte like cell line (U937), with the OP chlorpyrifos and found that chlorpyrifos induced the cell death of U937 in a dose- and time-dependent manner, as shown by 3-(4,5-Dimethylthiazol-2-yl)-2,5-diphenyltetrasolium bromide (MTT) and lactate dehydrogenase (LDH) assays and propidium iodide (PI) uptake. We then investigated whether chlorpyrifos-induced cell death consisted of apoptosis, as determined by a DNA fragmentation analysis and by analysis of Annexin-V staining and the intracellular level of active caspase-3 by flow cytometry. We found that chlorpyrifos induced apoptosis in U937 in a time- and dose-dependent manner, as shown by Annexin-V staining (Fig. 8). DNA fragmentation was detected when cells were treated with chlorpyrifos (Fig. 9). Chlorpyrifos also induced an increase in intracellular active caspase-3 in U937 cells in a dose-dependent manner, and a caspase-3 inhibitor, Z-DEVD-FMK, significantly inhibited the chlorpyrifos-induced apoptosis. These findings indicate that chlorpyrifos induced apoptosis in U937 cells [79]. Based on the above-mentioned findings, we further investigated whether OPs can induce apoptosis in human NK cells. We used NK-92CI and NK-92MI cells, which are interleukin-2 independent human NK cell lines, and that express CD56, perforin, granzymes A, B, 3/K and granulysin and are highly cytotoxic to K562 cells in the chromium release assay [50-52]. NK-92CI and/or NK-92MI were treated with DDVP or chlorpyrifos at 0-100ppm for 1-72 h at 37°C *in vitro*. Apoptosis induced by DDVP and chlorpyrifos was determined by FITC-Annexin V staining and the intracellular level of active caspase-3 was analyzed by flow cytometry. We found that DDVP and chlorpyrifos significantly induced apoptosis in NK-92CI and NK-92MI cells in a dose- and time-dependent manner (Fig.10). DDVP also induced an increase of intracellular active caspase-3 in NK-92CI in a dose- and time-dependent manner (Fig.11), and a caspase-3 inhibitor, Z-DEVD-FMK, significantly inhibited DDVP-induced apoptosis (Fig. 12), suggesting that this apoptosis is partially mediated by activation of intracellular caspase-3. The pattern of apoptosis induced by chlorpyrifos differed from that induced by DDVP. Chlorpyrifos showed a faster response than DDVP at higher doses; whereas, DDVP showed a slower, but stronger apoptosis-inducing ability than chlorpyrifos at lower doses (Fig. 13). We cannot explain the difference between the two OPs, but maybe the different water-solubility partially influences the apoptosis-inducing ability. Moreover, the response to OPs differed between NK-92CI and NK-92MI cells, and NK-92CI was more sensitive to OPs than NK-92MI (Fig. 14). This is similar to the inhibition of NK activity induced by DDVP, in which NK-92CI was more easily inhibited by DDVP than NK-92MI (Fig. 15), strongly suggesting a relationship between DDVP-induced apoptosis and the inhibition of cytolytic activity in NK cells. Taken together, these findings suggest that OP-induced inhibition of NK activity may be at least partially mediated by OP-induced apoptosis in NK [80].

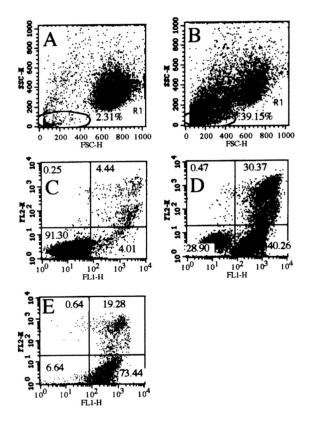

Figure 8. Chlorpyrifos-induced reduction in the cell size of U937. A: dot plot of FSC vs SSC in control U937 cells. B: dot plot of FSC vs SSC in chlorpyrifos-treated U937 cells: the horizontal axis (FSC) shows the size of U937 cell, while the vertical axis (SSC) shows the complexity. C: dot plot of FITC-Annexin V/PI in control U937 cells. D: dot plot of FITC-Annexin V/PI in chlorpyrifos-treated U937 cells. E: dot plot of FITC-Annexin V/PI in gate R1 in Figure 8B in chlorpyrifos-treated U937, the horizontal axis shows the intensity of fluorescence of FITC-Annexin V, whereas the vertical axis shows the intensity of fluorescence of PI. U937 cells were treated with chlorpyrifos at 0 or 142 μM for 4 hours, and then apoptosis was determined with FITC-Annexin V/PI staining detected with flow cytometry. Cited from Nakadai et al., Toxicology, 2006, 224, 202-209.

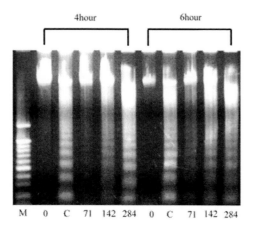

Figure 9. Chlorpyrifos-induced DNA fragmentation in U937 cells determined with agarose gel electrophoresis. M: marker of the DNA ladder, C: a positive control, camptotecin at 6 μM. The concentrations of chlorpyrifos were 0, 71, 142, and 284 μM. Data shown are representative of three similar experiments. Cited from Nakadai et al., Toxicology, 2006, 224, 202-209.

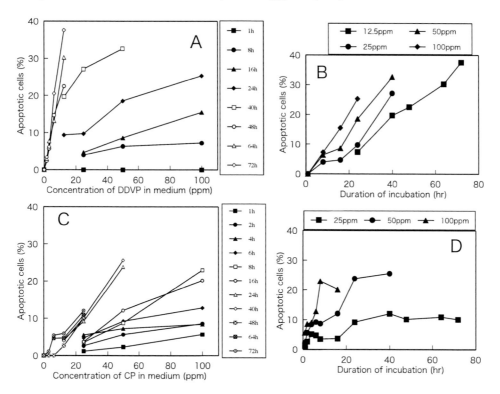

Figure 10. DDVP (A, B) and chlorpyrifos (CP) (C, D)-induced apoptosis in NK-92CI cells in dose-dependent (A, C) and time-dependent (B, D) manner determined by FITC-Annexin-V staining detected by flow cytometry. The correlation coefficients between the rate of apoptotic cells and the dose of OPs were r=0.994 (n=5, p<0.01) for DDVP at 72h, and r=0.990 (n=4, p=0.01) for CP at 8h; the correlation coefficients between the rate of apoptotic cells and the period of treatment were r=0.992 (n=5, p<0.01) for DDVP at 12.5ppm, and r=0.857 (n=11, p<0.01) for CP at 50ppm. Experiments were repeated at least 3 times with very similar results. Cited from Li et al., Toxicology 2007, 239, 89-95.

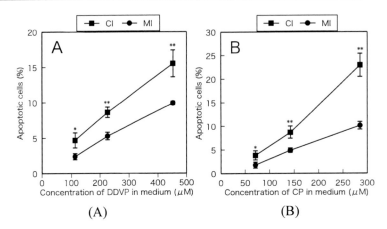

(A)   (B)

Figure 11. Different response to DDVP (A) and CP (B) in apoptosis between NK-92CI and NK-92MI cells. NK-92CI and NK-92MI cells were treated with DDVP for 16 hr or with CP for 8 hr. Data are presented as the mean ±SD (n=3). *: $p<0.05$, **: $p<0.01$, significantly different from NK-92MI cells by unpaired t-test. Cited from Li et al., Toxicology 2007, 239, 89-95.

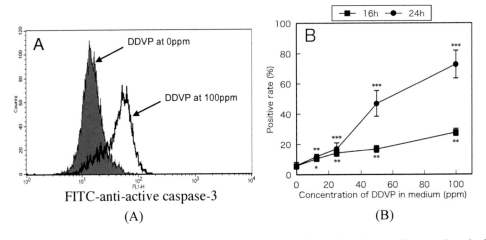

(A)   (B)

Figure 12. DDVP-induced increase in active caspase-3-positive NK-92CI cells. A: the shaded histogram shows the control cells (DDVP at 0ppm) and the open histograms show the cells treated with DDVP at 100ppm for 24 hr stained with FITC-rabbit anti-human active caspase 3. B: Data are presented as the mean ±SD (n=3 for 16h, n=5 for 24h). *: $p<0.05$, **: $p<0.01$, ***: $p<0.001$, significantly different from 0ppm by unpaired t-test. Cited from Li et al., Toxicology 2007, 239, 89-95.

Saleh et al. [81, 82] have also shown that paraoxon (the bioactive metabolite of parathion) and parathion cause apoptosis in a murine EL4 T-lymphocytic leukemia cell line through activation of caspase-3. In this study, pretreatment of EL4 cells with the caspase-9-specific inhibitor zLEHD-fmk attenuated paraoxon-induced apoptosis in a dose-dependent manner, whereas the caspase-8 inhibitor zIETD-fmk had no effect. Furthermore, activation of caspase-9, -8, and -3 in response to paraoxon treatment was completely inhibited in the presence of zLEHD-fmk, implicating the involvement of caspase 9-dependent mitochondrial pathways in paraoxon-induced apoptosis. Indeed, under both in vitro and in vivo conditions, paraoxon triggered a dose- and time-dependent translocation of cytochrome c from mitochondria into the cytosol and also disrupted the mitochondrial transmembrane potential.

Neither this effect nor cytchrome c release was dependent on caspase activation, since zVAD-fmk, the general inhibitor of the caspase family, did not influence either process. Finally, paraoxon treatment also resulted in a time-dependent up-regulation and translocation of the proapoptotic molecule Bax to mitochondria. Inhibition of this event by zVAD-fmk suggests that the activation and translocation of Bax to mitochondria follows activation of the caspase cascades. The results indicate that paraoxon induces apoptosis in EL4 cells through a direct effect on mitochondria by disrupting the transmembrane potential, causing the release of cytochrome c into the cytosol and subsequent activation of caspase-9 [82]. Das et al. [83] have also reported that OPs, such as monocrotophos, profenofos, chlorpyrifos, and acephate, significantly induced apoptosis and necrosis in cultured human peripheral blood lymphocytes in vitro using DNA diffusion assay.

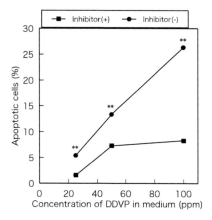

Figure 13. Effect of Z-DEVD-FMK on DDVP-induced apoptosis in NK cells using FITC-Annexin-V staining by flow cytometry. Data are presented as the mean ±SD (n=3). **: $p<0.001$, significantly different from the responses with inhibitor by unpaired t-test. Cited from Li et al., Toxicology 2007, 239, 89-95.

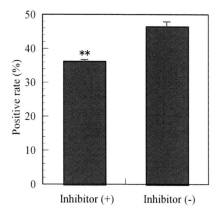

Figure 14. Effect of Z-DEVD-FMK on DDVP-induced increase of intracellular active caspase-3 in NK cells by flow cytometry. Data are presented as the mean ±SD (n=3). **: $p<0.001$, significantly different from the responses without inhibitor by unpaired t-test. Cited from Li et al., Toxicology 2007, 239, 89-95.

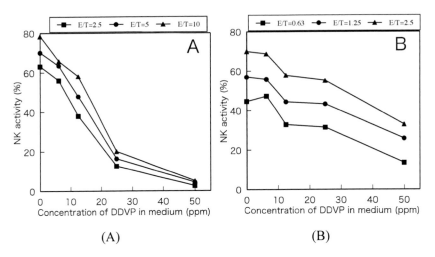

(A)   (B)

Figure 15. Different response to DDVP in cytolytic activity between NK-92CI (A) and NK-92MI (B) cells. Experiments were repeated at least 3 times with very similar results. Cited from Li et al., Toxicology 2007, 239, 89-95.

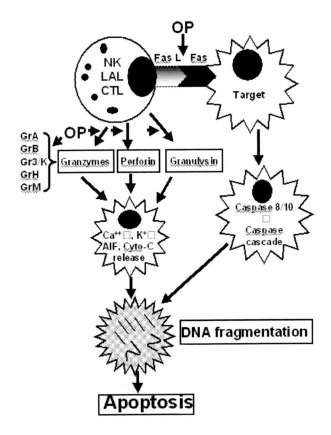

Figure 16. OP pesticides impair the granule exocytosis pathway (perforin/granzymes/granulysin pathway) and the FasL/Fas pathway of NK cells, LAK cells and CTLs. OP: organophosphorus pesticides. →: inhibition. Cited from Li and Kawada, Cellular and Molecular Immunlogy, 2006, 3, 171-178 and Li, JNMS, 2007, 74(2), 92-105.

In conclusion, the above findings indicate that OPs inhibit NK activitiy mediated by at least the following three mechanisms [58, 59]:

1). OPs impair the granule exocytosis pathway of NK cells (Fig. 16) [49-52,];
2). OPs impair the FasL/Fas pathway of NK cells (Fig. 16) [53];
3). OPs induce apoptosis of NK cells [80]. (Fig. 10)

Acknowledgements

This study was supported by grants from the Ministry of Education, Culture, Sports, Science and Technology of Japan (No. 09877077, No. 10770178, No. 12770206, No. 15590523 and No. 19590602). I am grateful to Professor Tomoyuki Kawada and other staff at the Department of Hygiene and Public Health, Nippon Medical School for their assistances and advice.

References

[1]   Costa, LG. Current issues in organophosphate toxicology. *Clin Chim Acta.* 2006, 366 (1-2), 1-13.

[2]   Richardson, ML. C309: Chlorpyrifos. In: Richardson ML, eds. *The dictionary of substances and their effects,* Vol. 2. Cambridge: The Royal Society of Chemistry; 1993; pp 460-463.

[3]   Richardson, ML. D257: Dichlorvos. In: Richardson ML, eds. *The dictionary of substances and their effects*, Vol. 3. Cambridge: The Royal Society of Chemistry; 1993; pp 391-396.

[4]   Ellenhorn, MJ; Barceloux, DG. Chapter 38: Pesticides. In: Ellenhorn MJ, Barceloux DG, eds. *Medical Toxicology.* New York: Elsevier; 1988; pp 1069-1077.

[5]   Japan Plant Protection Association. I. Production and Shipment of Pesticides. In: Japan Plant Protection Association, eds. *A handbook of pesticides.* Tokyo: Japan Plant Protection Association; 2003; pp 4-79 (in Japanese).

[6]   Pope, CN. Organophosphorus pesticides: do they all have the same mechanism of toxicity? *J Toxicol Envir Health*, Part B 1999, 2, 161-181.

[7]   Bajgar, J. Organophosphates/nerve agent poisoning: mechanism of action, diagnosis, prophylaxis, and treatment. *Adv Clin Chem* 2004, 38, 151-216.

[8]   Hermanowicz, A; Kossman, S. Neutrophil function and infectious disease in workers occupationally exposed to phosphoorganic pesticides: role of mononuclear-derived chemotactic factor for neutrophils. *Clin Immunol Immunopathol* 1984, 33, 13-22.

[9]   Rodgers, KE; Imamura, T; Devens, BH. Investigations into the mechanism of immunosuppression caused by acute treatment with O,O,S-trimethyl phosphorothioate. I. Characterization of the immune cell population affected. *Immunopharmacology* 1985, 10, 171-180.

[10]   Rodgers, KE; Ellefson, DD. Effects of acute administration of O,O,S-trimethyl phosphorothioate on the respiratory burst and phagocytic activity of splenic and peritoneal leukocytes. *Agents Actions* 1988, 24, 152-160.

[11]   Rodgers, KE; Ellefson, DD. Modulation of macrophage protease activity by acute administration of O,O,S trimethyl phosphorothioate. *Agents Actions* 1990, 29, 277-285.

[12]   Crittenden, PL; Carr, R; Pruett, SB. Immunotoxicological assessment of methyl parathion in female B6C3F1 mice. *J Toxicol Environ Health A* 1998, 54, 1-20.

[13]   Casale, GP; Cohen, SD; DiCapua, RA. The effects of organophosphate-induced cholinergic stimulation on the antibody response to sheep erythrocytes in inbred mice. *Toxicol Appl Pharmacol* 1983, 68, 198-205.

[14]   Johnson, VJ; Rosenberg, AM; Lee, K; Blakley, BR. Increased T-lymphocyte dependent antibody production in female SJL/J mice following exposure to commercial grade malathion. *Toxicology* 2002, 170, 119-129.

[15]   Pruett, SB; Chambers, JE. Effects of paraoxon, p-nitrophenol, phenyl saligenin cyclic phosphate, and phenol on the rat interleukin 2 system. *Toxicol Lett* 1988, 40, 11-20.

[16]   Casale, GP; Bavari, S; Connolly, JJ. Inhibition of human serum complement activity by diisopropylfluorophosphate and selected anticholinesterase insecticides. *Fundam Appl Toxicol* 1989, 12, 460-468.

[17]   Casale, GP; Vennerstrom, JL; Bavari, S; Wang TL. Inhibition of interleukin 2 driven proliferation of mouse CTLL2 cells, by selected carbamate and organophosphate insecticides and congeners of carbaryl. *Immunopharmacol Immunotoxicol* 1993, 15, 199-215.

[18]   Blakley, BR; Yole, MJ; Brousseau, P; Boermans, H; Fournier, M. Effect of chlorpyrifos on immune function in rats. *Vet Hum Toxicol* 1999, 41, 140-144.

[19]   Thrasher, JD; Madison, R;Broughton, A. Immunologic abnormalities in humans exposed to chlorpyrifos: preliminary observations. *Arch Environ Health* 1993, 48, 89-93.

[20]   Thrasher, JD; Heuser, G; Broughton A. Immunological abnormalities in humans chronically exposed to chlorpyrifos. *Arch Environ Health* 2002, 57, 181-187.

[21]   Youinou, P; Jamin, C; Pers, JO; Berthou, C; Saraux, A; Renaudineau, Y. B lymphocytes are required for development and treatment of autoimmune diseases. *Ann N Y Acad Sci* 2005, 1050, 19-33.

[22]   Rodgers, KE. Effects of oral administration of malathion on the course of disease in MRL-lpr mice. *J Autoimmun* 1997, 10, 367-373.

[23]   Ziem, G; McTamney, J. Profile of patients with chemical injury and sensitivity. *Environ Health Perspect* 1997, 105 Suppl 2, 417-436.

[24]   Berkson, J. Patient statement: a canary's tale. *Toxicol. Ind. Health* 1994, 10, 323-326.

[25]   Galloway, T; Handy, R. Immunotoxicity of organophosphorous pesticides. *Ecotoxicology* 2003, 12, 345-363.

[26]   Bertoncin, D; Russolo, A; Caroldi, S; Lotti, M. Neuropathy target esterase in human lymphocytes. *Arch Environ Health* 1985, 40, 139-144.

[27]   Handy, RD; Abd-El Samei, HA; Bayomy, MF; Mahran, AM; Abdeen, AM; El-Elaimy, EA. Chronic diazinon exposure: pathologies of spleen, thymus, blood cells, and lymph nodes are modulated by dietary protein or lipid in the mouse. *Toxicology* 2002,   172, 13-34.

[28]   Osicka-Koprowska, A; Lipska, M; Wysocka-Paruszewska, B. Effects of chlorfenvinphos on plasma corticosterone and aldosterone levels in rats. *Arch Toxicol* 1984, 55, 68-69.

[29]   Videira, RA; Antunes-Madeira, MC; Lopes, VI; Madeira, VM. Changes induced by malathion, methylparathion and parathion on membrane lipid physicochemical properties correlate with their toxicity. *Biochim Biophys Acta.* 2001, 1511, 360-368.

[30]   Li, Q; Minami, M; Clement, JG; Boulet, CA. Elevated frequency of sister chromatid exchanges in experiments by exposing lymphocytes to by-products generating of sarin synthesis:-Relating to Tokyo sarin disaster-. *Toxicol Lett* 1998, 98, 95-103.

[31]   Li, Q; Hirata, Y; Piao, S; Minami, M. The by-products generated during sarin synthesis in the Tokyo sarin disaster induced inhibition of natural killer and cytotoxic T lymphocyte activity. *Toxicology* 2000, 146, 209-220.

[32]   Hoppin, JA; Umbach, DM; London, SJ; Alavanja, MC; Sandler, DP. Chemical predictors of wheeze among farmer pesticide applicators in the Agricultural Health Study. *Am J Respir Crit Care Med* 2002, 165, 683–689.

[33]   Zhang, LX; Enarson,DA; He, GX; Li, B; Chan-Yeung, M. Occupational and environmental risk factors for respiratory symptoms in rural Beijing, China. *Eur Respir J* 2002, 20, 1525–1531.

[34]   Romagnani, S. Lymphokine production by human T cells in disease states. *Annu Rev Immunol* 1994, 12, 227–257.

[35]   de Vries, E; de Bruin-Versteeg, S; Comans-Bitter, WM; de Groot, R; Hop, WC; Boerma, GJ, et al. Longitudinal survey of lymphocyte subpopulations in the first year of life. *Pediatr Res* 2000, 47, 528–537.

[36]   Kuo, ML; Huang, JL; Yeh, KW; Li, PS; Hsieh, KH. Evaluation of Th1/Th2 ratio and cytokine production profile during acute exacerbation and convalescence in asthmatic children. *Ann Allergy Asthma Immunol* 2001, 86, 272–276.

[37]   Yssel, H; Abbal, C; Pene, J; Bousquet, J. The role of IgE in asthma. *Clin Exp Allergy* 1998, 28(suppl 5), 104–109.

[38]   Heaton, T; Rowe, J; Turner, S; Aalberse, RC; de Klerk, N; Suriyaarachchi, D, et al. An immunoepidemiological approach to asthma: identification of in-vitro T-cell response patterns associated with different wheezing phenotypes in children. *Lancet* 2005, 365, 142–149.

[39]   Cohn, L; Elias, JA; Chupp, GL. Asthma: mechanisms of disease persistence and progression. *Annu Rev Immunol* 2004, 22, 789–815.

[40]   Sanchez-Guerrero, I; Vegara, RP; Herrero, N; Garcia-Alonso, AM; Luna, A; Alvarez, MR. Cytokine serum profiles in allergic and non-allergic asthma. Increased production of IL-10 by non-allergic asthmatic patients. *Allergol Immunopathol (Madr)* 1997, 25, 98–103.

[41]   Duramad, P; Tager, IB; Leikauf, J; Eskenazi, B; Holland, NT. Expression of Th1/Th2 cytokines in human blood after in vitro treatment with chlorpyrifos, and its metabolites, in combination with endotoxin LPS and allergen Der p1. *J Appl Toxicol* 2006, 26, 458-65.

[42]   Duramad, P; Harley, K; Lipsett, M; Bradman, A; Eskenazi, B; Holland, NT; Tager, IB. Early environmental exposures and intracellular Th1/Th2 cytokine profiles in 24-month-old children living in an agricultural area. *Environ Health Perspect* 2006, 114, 1916-1922.

[43] Masuda, N; Takatsu, M; Morinari, H; Ozawa, T. Sarin poisoning in Tokyo subway. *Lancet* 1995, 345, 1446.

[44] Richardson, ML; Gangolli S. S7: Sarin. In: Richardson ML, Gangolli S, eds. *The dictionary of substances and their effects,* Vol. 7. Cambridge: The Royal Society of Chemistry; 1993; pp 10-11.

[45] Li, Q; Hirata, Y; Kawada, T; Minami, M. Elevated frequency of sister chromatid exchanges of lymphocytes in sarin-exposed victims of the Tokyo sarin disaster 3 years after the event. *Toxicology* 2004, 201, 209-217.

[46] Li, Q; Minami, M. Sister chromatid exchanges of human peripheral blood lymphocytes induced by N, N-diethylaniline *in vitro. Mutation Res* 1997, 395, 151-157.

[47] Li, Q; Hirata, Y; Piao, S; Minami, M. Immunotoxicity of N, N-diethylaniline in mice: Effect on natural killer activity, cytotoxic T lymphocyte activity, lymphocyte proliferation response and cellular components of the spleen. *Toxicology* 2000, 150, 181-191.

[48] Minami, M; Hui, D-M; Wang, Z; et al. Biological monitoring of metabolites of sarin and its by-products in human urine samples. *J Toxicol Sci* 1998, 23 Suppl 2, 250-254.

[49] Li, Q; Nagahara, N; Takahashi, H; Takeda, K; Okumura, K; Minami, M. Organophosphorus pesticides markedly inhibit the activities of natural killer, cytotoxic T lymphocyte and lymphokine-activated killer: a proposed inhibiting mechanism via granzyme inhibition. *Toxicology* 2002, 172, 181-190.

[50] Li, Q; Nakadai, A; Ishizaki, M; Morimoto, K; Ueda, A; Krensky, AM; Kawada, T. Dimethyl 2,2-dichlorovinyl phosphate (DDVP) markedly decreases the expression of perforin, granzyme A and granulysin in human NK-92CI cell line. *Toxicology* 2005, 213, 107-116.

[51] Li, Q; Nakadai, A; Matsushima, H; Miyazaki, Y; Krensky, AM; Kawada, T; Morimoto, K. Phytoncides (wood essential oils) induce human natural killer cell activity. *Immunopharmacol Immunotoxicol.* 2006, 28, 319-333.

[52] Li, Q; Kobayashi, M; Kawada T. DDVP markedly decreases the expression of granzyme B and granzyme 3/K in human NK cells. 2008, 243, 294-302.

[53] Li, Q; Nakadai, A; Takeda, K; Kawada, T. Dimethyl 2,2-dichlorovinyl phosphate (DDVP) markedly inhibits activities of natural killer cells, cytotoxic T lymphocytes and lymphokine-activated killer cells via the Fas-ligand/Fas pathway in perforin-knockout (PKO) mice. *Toxicology* 2004, 204, 41-50.

[54] Zabrodskii, PF; Germanchuk, VG. Role of activation of the sympathoadrenal system in the realization of immune reactions during acute poisoning with organophosphorus compounds. *Bull Exp Biol Med.* 2001, 132, 966-968.

[55] Rodgers, KE; Grayson, MH; Ware, CF. Inhibition of cytotoxic T lymphocyte and natural killer cell-mediated lysis by O,S,S,-trimethyl phosphorodithioate is at an early postrecognition step. *J Immunol* 1988, 140, 564-570.

[56] Rodgers, KE; Leung, N; Ware, CF. Effects of acute administration of O,S,S-trimethyl phosphorodithioate on the generation of cellular and humoral immune responses following in vitro stimulation. *Toxicology* 1988, 51, 241-253.

[57] Rodgers, KE; Stern, ML; Ware, CF. Effects of subacute administration of O,S,S-trimethyl phosphorodithioate on cellular and humoral immune response systems. *Toxicology* 1989, 54, 183-195.

[58]   Li, Q; Kawada T. The new mechanism of organophosphorus pesticides-induced inhibition of cytolytic activity of killer cells. *Cellular & Molecular Immunology (CMI)* 2006, 3, 171-178.

[59]   Li, Q. New mechanism of organophosphorus pesticide-induced immunotoxicity. *J Nippon Med Sch.* 2007, 74, 92-105.

[60]   Kagi, D; Vignaux, F; Ledermann, B; et al. Fas and perforin pathways as major mechanisms of T cell-mediated cytotoxicity. *Science* 1994, 265, 528-530.

[61]   Mori, S; Jewett, A; Murakami-Mori, K; Cavalcanti, M; Bonavida, B. The participation of the Fas-mediated cytotoxic pathway by natural killer cells is tumor-cell-dependent. Cancer *Immunol Immunother* 1997, 44, 282-290.

[62]   Sayers, TJ; Brooks, AD; Lee, JK; et al. Molecular mechanisms of immune-mediated lysis of murine renal cancer: differential contributions of perforin-dependent versus Fas-mediated pathways in lysis by NK and T cells. *J Immunol* 1998, 161, 3957-3965.

[63]   Smyth, MJ; Trapani, JA. Granzymes: exogenous proteinases that induce target cell apoptosis. *Immunol Today* 1995, 16, 202-206.

[64]   Okada, S; Li, Q; Whitin, JC; Clayberger, C; Krensky, AM. Intracellular mediators of granulysin-induced cell death. *J Immunol* 2003, 171, 2556-2562.

[65]   Kagi, D; Ledermann, B; Burki, K; et al. Cytotoxicity mediated by T cells and natural killer cells is greatly impaired in perforin-deficient mice. *Nature* 1994, 369, 31-37.

[66]   Gershenfeld, HK; Hershberger, RJ; Shows, TB; Weissman, IL. Cloning and chromosomal assignment of a human cDNA encoding a T cell- and natural killer cell-specific trypsin-like serine protease. *Proc Natl Acad Sci USA* 1988, 85, 1184-1188.

[67]   Sayers, TJ; Brooks, AD; Ward, JM; et al. The Restricted Expression of Granzyme M in Human Lymphocytes. *J Immunol* 2001, 166, 765-771.

[68]   Trapani, JA; Klein, JL; White, PC; Dupont, B. Molecular cloning of an inducible serine esterase gene from human cytotoxic lymphocytes. *Proc Natl Acad Sci USA* 1988, 85, 6924-6928.

[69]   Sayers, TJ; Lloyd, AR; McVicar, DW; et al. Cloning and expression of a second human natural killer cell granule tryptase, HNK-Tryp-2/granzyme 3. *J Leukoc Biol* 1996, 59, 763-768.

[70]   Hirata, Y; Inagaki, H; Shimizu, T; Li, Q; Nagahara, N; Minami, M; Kawada, T. Expression of enzymatically active human granzyme 3 in Escherichia coli for analysis of its substrate specificity. *Arch Biochem Biophys* 2006, 446, 35-43.

[71]   Meier, M; Kwong, PC; Fregeau, CJ; et al. Cloning of a gene that encodes a new member of the human cytotoxic cell protease family. *Biochemistry* 1990, 29, 4042-4049.

[72]   Smyth, MJ; Sayers, TJ; Wiltrout, T; Powers, JC; Trapani, JA. Met-ase: cloning and distinct chromosomal location of a serine protease preferentially expressed in human natural killer cells. *J Immunol* 1993, 151, 6195-6205.

[73]   Mahrus, S; Craik, CS. Selective chemical functional probes of granzymes A and B reveal granzyme B is a major effector of natural killer cell-mediated lysis of target cells.
Chem Biol. 2005, 12(5), 567-77.

[74]   Hudig, D; Redelman, D; Minning, LL. The requirement for proteinase activity for human lymphocyte-mediated natural cytotoxicity (NK): evidence that the proteinase

is serine dependent and has aromatic amino acid specificity of cleavage. *J Immunol.* 1984, 133(5), 2647-54.

[75]   Liu, CC; Walsh, CM; Eto, N; Clark, WR; Young, JD. Morphologic and functional characterization of perforin-deficient lymphokine-activated killer cells. *J Immunol* 1995, 155, 602-608.

[76]   Caughlan, A; Newhouse, K; Namgung, U; Xia, Z. Chlorpyrifos induces apoptosis in rat cortical neurons that is regulated by a balance between p38 and ERK/JNK MAP kinases. *Toxicol Sci* 2004, 78, 125-134.

[77]   Carlson, K; Jortner, BS; Ehrich, M. Organophosphorus compound-induced apoptosis in SH-SY5Y human neuroblastoma cells. *Toxicol Appl Pharmacol* 2000, 168, 102-113.

[78]   Greenlee, AR; Ellis, TM; Berg, RL. Low-dose agrochemicals and lawn-care pesticides induce developmental toxicity in murine preimplantation embryos. *Environ Health Perspect* 2004, 112, 703-709.

[79]   Nakadai, A; Li, Q; Kawada, T. Chlorpyrifos induces apoptosis in human monocyte cell line U937. *Toxicology* 2006, 224, 202-209.

[80]   Li, Q; Kobayashi, M; Kawada, T. Organophosphorus pesticides induce apoptosis in human NK cells. *Toxicology* 2007, 239, 89-95.

[81]   Saleh, AM; Vijayasarathy, C; Masoud, L; Kumar, L; Shahin, A; Kambal, A. Paraoxon induces apoptosis in EL4 cells via activation of mitochondrial pathways. *Toxicol Appl Pharmacol* 2003, 190, 47-57.

[82]   Saleh, AM; Vijayasarathy, C; Fernandez-Cabezudo, M; Taleb, M; Petroianu, G. Influence of paraoxon (POX) and parathion (PAT) on apoptosis: a possible mechanism for toxicity in low-dose exposure. *J Appl Toxicol* 2003, 23, 23-29.

[83]   Das, GP; Shaik, AP; Jamil, K. Estimation of apoptosis and necrosis caused by pesticides in vitro on human lymphocytes using DNA diffusion assay. *Drug Chem Toxicol* 2006, 29, 147-156.

Chapter 4

NKT Cells: Dual Role in Tumor Immunosurveillance and Implications in Cancer Immunotherapy

Sergiy V. Olishevsky[*], *Alexandra L. Sevko*[**] *and Volodymyr O. Shlyakhovenko*

R.E. Kavetsky Institute of Experimental Pathology,
Oncology and Radiobiology of National Academy of Sciences of Ukraine
Vasylkivska Street 45, 03022 Kyiv, Ukraine.

Abstract

Natural killer T (NKT) cells are one of the most mysterious immune system cells represented by unusual regulatory T lymphocytes which co-express some NK cell markers, and have the capacity to recognize glycolipid antigens in the context of MHC I-like molecule – CD1d via their invariant T cell receptor (TCR). In response to TCR ligation, NKT cells rapidly produce large amounts of both pro-inflammatory T helper (Th) 1 type cytokines and anti-inflammatory Th2 type cytokines. This paradoxical property has made it difficult to predict the consequences of NKT cell activation *in vivo* but has nonetheless caused much speculation that NKT cells play a grand role in immunoregulation.

The physiological function of NKT cells in antitumor immunity may be multifaceted. Despite accumulated experimental evidence that has supported the critical role of NKT cells in promoting effective tumor immunosurveillance, several studies suggest a contrary role for NKT cells consisted in dramatic suppression of acquired antitumor immunity. Recent studies discovered a veiled enigma of controversial role of NKT cells for tumor immunosurveillance and revealed that at least two subsets of CD1d-restricted NKT cells exist: type I NKT cells that can substantially influence function of other various cell types, particularly DC, NK cells, conventional CD4[+] and CD8[+] T cells,

---

[*] E-mail address: sergeyolishevsky@yahoo.com
[**] E-mail address: sevko@ukr.net

all contributing to the antitumor immunity, and type II NKT cells that have a potency to repress antitumor immune responses.

Immunotherapy strategies aimed at augmentation of CD1d-restricted type I NKT cell numbers or increasing of their activity are currently a major focus. Many latest experimental data and clinical trials clearly suggested that targeting NKT cells may provide a novel effective strategy for immunotherapy of incurable patients with malignancies; moreover frequency and/or function of NKT cells may be directly related to cancer disease prognosis.

Introduction

The last two decades have seen an appreciable renaissance of interest in tumor immunosurveillance and a broadening of this concept into one termed tumor immunoediting [8, 27]. Immunosurveillance refers to the policing or monitoring function of immune system cells to recognize and eliminate clones of transformed cells prior to their development into neoplasms and to destroy developed tumors. Rapidly accumulating data have begun to elucidate the cellular basis of tumor immunosurveillance and demonstrate that lymphocytes of both the adaptive and innate immunity can effectively prevent tumor development. However, recent works have shown that the immune system may also promote the emergence of primary tumors with reduced immunogenicity that are capable to escape immune recognition [107]. Natural killer T (NKT) cells, that are in the main focus of this review, participating in tumor immunosurveillance represent a striking paradigm of "double-edged sword" in tumor immunity regulation.

NKT cells were first described in 1987 as murine thymocytes that express a restricted T cell antigen (Ag) receptor (TCR) αβ repertoire in combination with the NK cell marker, C-type lectin NK1.1 (CD161) [30]. In fact, this unusual cell population was found and characterized by three independent lines of research [5, 30, 57]. Interestingly, NKT cells constitute the unique T lymphocyte subpopulation that shares several characteristics with NK cells and mediate potent immunoregulatory effects *in vivo*. NKT cells recognize glycolipid Ags in the context of the monomorphic major histocompatibility complex-like molecule, CD1d (reviewed in [145]). Like conventional T cells, NKT cells develop from CD4□ CD8□ thymic progenitor T cells (reviewed in [67]). Immature $CD4^+CD8^+$ T cells arising from these progenitor cells develop further into the NKT cell lineage upon appropriate signaling by the monomorphic CD1d Ag-presenting molecule that is expressed by cortical thymocytes [7, 98]. It is important to note that in the absence of the CD1d molecule, NKT cells do not develop [71]. Additionally, endosomal trafficking of CD1d in Ag-presenting cells (APC) plays an important role in NKT cell development, presumably due to the necessity of CD1d-loading with endogenous Ag during NKT cell selection [18].

Subsequent investigations found that the vast majority of NKT cells are either $CD4^+$ or $CD4^+CD8^+$ and utilize an invariant TCR α-chain rearrangement [6, 60, 128, 129]. The emerging evidence has demonstrated that, among the CD1d-restricted NKT cells, at least two subtypes exist [39]. Type I NKT cells, also called invariant NKT (iNKT) cells, have a highly restricted TCR repertoire and exclusively express an invariant Ag receptor Va14Ja281 in mice and Va24JaQ in human [24, 29, 58]. Therefore, Va14/Va24 Ag receptor can be considered as an exclusive marker for type I NKT cells. CD1d-restricted type II NKT cells, which are referred to non-iNKT cells, because express a more diverse TCR repertoire [39, 145].

NKT cells have a unique capacity to rapidly produce large amounts of both T helper (Th) 1 and Th2 cytokines, including interferon (IFN)-γ, interleukin (IL)-4, tumor necrosis factor (TNF)-α, granulocyte-macrophage colony-stimulating factor (GM-CSF), and IL-13, immediately after Ag stimulation by retaining a constitutively high level of cytokine mRNAs [68, 123]. In due time, this paradoxical capacity of NKT cells led to many speculations and discussions on their very important role in immunoregulation. Moreover, a crucial role of NKT cells in various immune responses was also hypothesized because of the observation that CD1d/NKT cell system was highly conserved through evolution [121]. A number of studies have diligently analyzed the role of this mysterious T cell with NK cell phenotype in control of autoimmunity (reviewed in [4, 44, 76]), infectious diseases (reviewed in [9, 41]), allergy and inflammation (reviewed in [19, 74, 102]), and, finally, tumor immunosurveillance (reviewed in [14, 104]). Evidence for NKT cell participation in these immune responses has been established and is intensively investigated by many laboratories in the world.

Interestingly, NKT cells are relatively resistant to apoptosis and it is possible that their anti-apoptotic properties are linked to the regulatory role of the NKT cell population [103]. Furthermore, recent evidence is also included that, unlike type I NKT cells, type II NKT cells have the capacity to negatively affect specific antitumor immune responses and substantially contribute to repression of antitumor immunity [38, 39, 104, 118].

Since, most studies NKT cells have focused on iNKT cells, therefore in this chapter, unless otherwise indicated, the term "NKT cell" refers to the type I NKT cells. In this review, we will highlight recently accumulated knowledge of the principal mechanisms by which NKT cells recognize and respond to tumors in physiological and immunotherapeutic settings. The recent advances in understanding the importance of NKT cells for tumor immunosurveillance regulation will also reviewed here. In addition, we will discuss results of clinical trials with implication of NKT cell-based immunotherapeutic strategies and define the future prospects in antitumor NKT cell biology and cancer immunotherapy.

NKT Cells are the One of the Main Players in Tumor Immunosurveillance

The potential importance of NKT cells in antitumor immunity has been well established. Accumulated experimental evidence supports that NKT cells can directly or via other immune system cells or cytokines participate in elimination of cancer cells or inhibit an expansion of cancer cell growth directly or via other immune system cells or cytokines [14, 59, 104].

Initially, a role for IL-4-producing NK1.1$^+$CD4$^-$CD8$^-$ T cells was suggested in the study with rejection of embryonal carcinoma cells [84]. More recently, experiments using NKT cell-deficient Jα28$^{-/-}$ mice confirmed a critical role for NKT cells in protection from spontaneous sarcomas initiated by a chemical carcinogen, methylcholantren [119]. Further evaluation data obtained using methylcholanthren-induced tumor models, in conjunction with NKT cell-transfer experiments [115], and similar findings reported for experimental model of pulmonary sarcoma metastasis [94] finally revealed that NKT cells play an important role in protection against tumor growth.

It has been demonstrated that activated NKT cells, like NK and cytotoxic T cells, induce tumor cell death by the expression of a wide variety of cell-death-inducing effector molecules, including perforin, TNF-α, Fas ligand, or TRAIL [52, 83, 86]. However, when mechanisms of NKT cell anitumor activity have been studied in detail, it appeared that these potentially tumoricidal products are dispensable and NKT cell-derived IFN-γ was shown to be critical in NKT-mediated antitumor immunity [46, 117]. These studies have suggested that NKT cells are not acting primarily as effectors, but in first turn they recruit and promote activity of other immunocompetent cells in IFN-γ-dependent manner. Now it is well known that stimulated NKT cells have a remarkable capacity to produce both Th1 (e.g., IFN-γ, TNF-α, GM-CSF) and Th2 (e.g., IL-4, IL-10) cytokines, which can subsequently contribute to activation of various immune system cells, such as dendritic cells (DCs), NK cells, or cytotoxic T cells (Fig. 1) [39, 104, 116]. Therefore, NKT cells are believed to mediate direct cytotoxicity and also exhibit the "adjuvant effects" on antitumor immunity by activating other cytotoxic cells mainly through Th1 cytokine cascades [104].

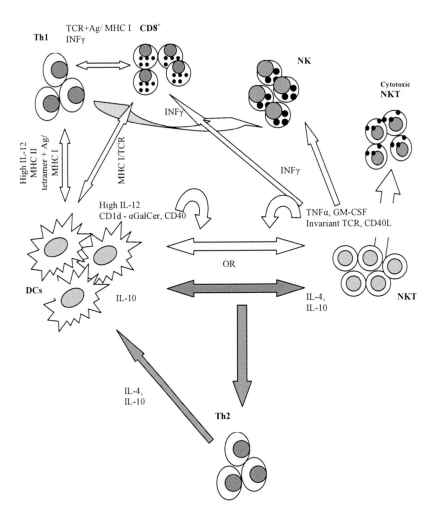

Figure 1. DCs and NKT cells are the key cells to regulate immune response. Interaction between DCs and NKT cells supplies Th1/Th2 balance.

Based on recently accumulated knowledge of the antitumor function of NKT cells, Seino *et al.* [104] have proposed two theories explaining possible mechanisms of NKT cell involvement to cancer immunosurveillance. The first theory consists in NKT cells might be able to respond to any type of tumor cells irrelevant of the tumor specificity because NKT cells might mediate antitumor responses without direct recognition of tumor-derived Ags. NKT cells receive a basal level of stimulation by endogenous ligands (likely glycolipids) presented by CD1d through their invariant TCR. In addition, NKT cells can become activated via APC stimulated during a variety of infections and inflammatory responses [10, 11, 38, 61, 129, 146]. As far antitumor responses sometimes associate with inflammatory reactions that deliver cytokines such as IL-12, and during the tumor-mediated inflammatory processes it is produced, NKT cells can be activated by it. The recent identification of bacterial origin Ags for NKT cells provides an explanation for the capacity of certain microorganisms to activate NKT cells. In the case of microbial infection, NKT cells also seem to be activated IL-12 producing DC stimulated via Toll-like receptors in the presence of a basal level of a CD1d/TCR-mediated signal [11].

The second theory assumes that it is still possible that NKT cells may directly recognize glycolipid components from tumor cells. In fact, some glycolipid fractions from the tumor cell membrane are known to be presented by CD1d and to be recognized by NKT cells [43]. Furthermore, an endogenous glycolipid ligand for NKT cells has been recently identified as an isoGb3 [154], that can stimulate NKT cells in a way similar to nature exogenous ligand α-galactosylceramide (α-GalCer) [152]. Thus, it is possible that NKT cells can induce antitumor responses as a result of their recognition of similar to isoGb3 tumor cell-derived glycolipid Ags.

However, recent reports have shown that in spite of possibility of NKT cell-mediated lysis of CD1d-expressing tumor cells [72], some tumors may escape NKT-mediated direct cytotoxicity by through shedding of neutral glycolipids [122] that could lead to altered NKT activation and consequent suppression of antitumor response [39, 61].

Antitumor Activities of Resident or Stimulated Mouse and Human NKT Cells

Several years ago, two main subpopulations of I type of mouse NKT cells have been identified: CD4$^+$ and CD4$^-$CD8$^-$ NKT cells. However, functional differences between them have remained unclear [105]. In a recent study, Crowe *et al.* compared antitumor capacities of distinct subsets of NKT cells isolated from different organs in mice [20]. NKT cells from liver, thymus, and spleen were compared for their ability to mediate rejection of two different types of tumor cells (methylcholanthrene-induced sarcoma cells and B16 melanoma cells) *in vivo*. Interestingly, it was found that the CD4$^-$CD8$^-$ NKT cells in liver showed superior antitumor activities than CD4$^+$ NKT cells from liver and both the CD4$^+$ and CD4$^-$ CD8$^-$ NKT cell subsets originated from thymus and spleen. This study suggests that organ-specific mechanisms might dictate the functional capabilities of resident NKT cells. Indeed, several past studies have presented evidence that NKT cells in other organs may also differ in functions (reviewed in [105]).

Stimulated with α-GalCer liver-derived NKT cells were also more able to reject B16 melanoma metastases as compared with NKT cells isolated from other organs [20]. However, the mechanism of these antitumor effects of NKT cells remains uncertain, because Crowe and colleagues did not found significant differences in cytokine production (IL-4 and IFN-γ) by CD4$^+$ and CD4$^-$CD8$^-$ NKT cells isolated from different organs.

In human, the frequency of Vα24 NKT cells is very low. In spite of this, the human Vα24 NKT cell lines expanded by the stimulation of peripheral blood mononuclear cells with α-GalCer *in vitro* show substantial cytotoxicity against different tumor target cells [25, 53, 72, 127]. Furthermore, human NKT cells stimulated *in vitro* with α-GalCer also can activate NK cells in IL-2-dependent manner [72] that highlights the potential of NKT cell stimulation for immunotherapy of cancer patients.

In their report, van der Vliet *et al.* [144] have shown that freshly isolated and expanded Vα24 NKT cells express granzyme B but not FasL. In contrast, Gumperz *et al.* [42] found that CD4$^+$ human NKT cells after phorbol myristate acetate or ionomycin stimulation demonstrate a high expression of FasL, while CD4$^-$ NKT cells selectively produce TNF-α and express perforin. It has also demonstrated that Vα24 NKT cells possess cytoplasmic perforin and kill U937 cells mainly through a perforin-mediated pathway [53]. Moreover, human NKT cells can express TRAIL, thus inducing apoptosis in TRAIL-sensitive leukemic cells [86].

Seino *et al.* [104] have pointed that the various antitumor activities of human NKT cells can be probably explained by the use of distinct cytotoxic machinery adjusted to the microenvironment and the sensitivity of target cells.

Tumor Immunosurveillance Suppression Induced by Type II NKT Cells

One of numerous mechanisms of cancer immune evasion can be realize by the suppression of antitumor immunity with immunoregulatory T cells. Since the concept of "suppressor" or "regulatory" T cells was reintroduced to the immunology community, a large massive of knowledge has been accumulated regarding several regulatory T-cell populations, their roles and mechanisms of action. Such cells were first re-discovered in autoimmune disease. As cancers develop as a part of "self", such regulatory T cells also might regulate antitumor immune responses [132].

Indeed, there has been a rapid growth in the number of reports describing potent immunoregulatory role of NKT cells in tumor immunosurveillance. In the context of tumor immunology, NKT cells have been also shown to suppress antitumor immunity as well [59, 78, 94, 119, 133]. Immunosuppressive CD4$^+$ NK cells are attributed to type II NKT cells and their function *in vivo* has been investigated only by comparing results from experiments on Jα281- and CD1d-deficient mouse models. It is important to note that in Jα281-deficient mice, only type I NKT cells are lacking, [21], but in CD1d-deficient mice, both type I and II NKT cells are absent [71].

In 2000, two pioneering independent papers were published in *Nature Immunology* (Vol. 1, no. 6) and highlighted previously unrecognized mechanisms that regulate cell-mediated antitumor immunity. Terabe *et al.* [133] provided compelling data that defines a new function for CD4$^+$ NKT cells as potential suppressors of cytotoxic T cell-mediated antitumor immunity. Other related study was originally performed by Moodycliffe et al. [78] and showed that in ultraviolet (UV)-induced tumor model, CD4$^+$ T cells that express NK cell marker, DX5, play a critical role in immunosuppression. Similar observations also were obtained earlier Matsui et al. [69] and indicated that CD4$^+$CD25$^-$ T cells suppressed CD8$^+$ T cells to prevent complete elimination of tumors in a transformed fibrosarcoma model.

These two studies were the first successful steps towards full understanding the immunoregulatory activity of NKT cells and their immunosuppressive potential for cancer immunosurveillance. First, Terabe et al. [133], using a model of tumor recurrence, have discovered that CD1d-restricted NKT cells through an activated IL-4Rα-STAT6 signaling pathway specifically prevent effective T killer cell-mediated tumor eradication in an IL-13-dependent manner. Thus, it was the first study to demonstrate a key role for NKT cells, and for IL-13, in the repression of T cell-mediated antitumor immunosurveillance. Second, Moodycliffe et al. [78] identified CD4$^+$DX5$^+$ cells which represent a distinct from CD4$^+$NK1.1$^+$ cells regulatory population of NKT cells and are the key repressor cells in a well-established model of UV-induced immunosuppression. So, this discovery supported by earlier findings [44], has provided evidence of the possibility of NKT cell heterogeneity that could explain a potential functional diversity of T cells that carry NK markers.

Finally, Smyth and Godfrey [118] in their critical paper, also published in this issue of *Nature Immunology,* have discussed in detail the immunosuppressive potential of NKT cells as "the other side of the coin" of tumor immunosurveillance regulation.

However, the new problem on the way of explanation of the negative regulation of tumor immunosurvailance by NKT cell was that cytotoxic T lymphocytes do not express IL-13 receptors and therefore can not be directly activated with IL-13 produced by type II NKT cells. An elegant study of Terabe et al. [132, 134] subsequently demonstrated that NKT cell-derived IL-13 activates CD11b$^+$Gr-1$^+$ myeloid cells to produce TGF-β, which directly suppresses antitumor activity of cytotoxic CD8$^+$ T lymphocytes. They showed that blocking of TGF-β or depletion of Gr-1$^+$ cells *in vivo* completely prevented tumor recurrence. Thus, TGF-β was found to responsible for negative regulation antitumor immunity by II type NKT cells. This negative regulatory circuit was also found in the study with lung metastasis model of the murine carcinoma [135].

In more recent work, Terabe et al. [136] finally identified type II NKT cells as the regulatory cells involved in the tumor immunity suppression. They compared the negative regulation of effects observed in Jα281-deficient mice that lack type I NKT cells with those in CD1d-deficient mice that lack both type I and type II NKT cells. The study showed that CD1d-deficient mice but not Jα281-deficient mice failed to suppress antitumor responses.

Regulation of NKT Cell-Mediated Antitumor Responses

It is unclear whether the different subsets or types of NKT cells regulate each other's function. We discussed above the study of Crowe et al. [20] which have found significant differences in antitumor activity of different NKT cell subsets isolated from different organs and attempted to elucidate possible mechanism implicated to this phenomenon. This study demonstrated that thymic cells from IL4-deficient mice acquire the ability to reject tumors, suggesting that the production of IL-4 might inhibit certain antitumor responses *in vivo*. However, molecules other than IL-4 must also contribute to the inhibition of NKT cell-mediated immunity, as hepatic CD4⁻CD8⁻ NKT cells that mediated tumor rejection in this model were shown to produce significant levels of IL-4 [20]. Furthermore, Seino and Taniguchi [105], analyzing this and other similar reports, have suggested that antitumor function of liver-derived CD4⁻CD8⁻ NKT cells is controlled by other subsets or types of NKT cells, in a manner similar to the regulatory interplay between Th1 and Th2 cells.

Several recent papers indicated that immunoregulatory CD4⁺CD25⁺ T cells (Treg) can suppress NKT cell antitumor effects. In fact, Azuma *et al.* [3] reported that Treg can suppress the proliferation, cytokine production (IFN-γ, IL-4, IL-13 and IL-10) and cytotoxic activity of the CD4⁺ and CD4⁻CD8⁻ subsets of NKT cells. The mechanism of NKT-cell suppression by Treg was similar to the mechanism by which these cells suppress activity of conventional T and B cells. Specifically, the suppression of NKT-cell responses required cell-cell contact with the NKT cells and was independent of soluble factors [63]. This was revealed by studies showing that suppression was blocked by anti-ICAM-1 antibodies or when cells were separated by transwell membranes, whereas suppression was maintained following the addition of neutralizing antibodies against IL-10 and TGF-β [3].

Furthermore, Nishikawa *et al.* [88] also found that Treg cells can suppress the antitumor activity of NKT cells, at least in certain experimental systems. These studies used tumor models in which NKT cells were primarily responsible for tumor suppression. In brief, mice were immunized with plasmids encoding some autoAgs before tumor challenge that resulted in the marked enhancement of tumor growth. Genetic immunization induced Tregs that could down-regulate the numbers and activity of NKT cells, resulting in tumor growth enhancement.

Additionally, the recent evidence showed a reciprocal cross-talk between Treg and NKT cells that should be extremely important for deeper understanding of immunoregulation machinery in health organism and in disease statues, including cancer [63].

NKT Cells in Cancer Patients: Quantitative and Qualitative Alterations and Clinical Prognosis

Quantitative and qualitative deficiencies in the NKT cells have been reported in association with majority, but not all, types of human cancer. NKT cell alterations were observed in patients with colon cancer, lung cancer, breast cancer, melanoma, head and neck squamous cell carcinoma, prostate cancer, myelodysplastic syndromes, and progressive malignant melanoma, but were not in glioma patients (reviewed in [142]). Clinical studies as well as experimental investigations declared that NKT cells from various cancer patients show a marked reduction in the size of the NKT cell population [104, 131]. The most significant number reduction of NKT cells was registered in peripheral blood of patients with advanced lung cancer [80]. Accumulating evidence in humans indicates that numerous defects of the NKT pool are indeed of clinical significance, as they are associated with poor prognosis in patients with neuroblastoma and colon cancer [73, 125].

The quantitative defects in NKT numbers in many cancer patients are accompanied by qualitative defects in the residual NKT population, including significant decrease of its ability to produce cytokines, or proliferate even in response to stimulation with α-GalCer as compared to cells from healthy individuals [25, 54, 126, 143, 149]. Indeed, in prostate cancer patients, NKT cells were markedly decreased in number and showed diminished *ex vivo* expansion when stimulated with the a-GalCer. Furthemore, the amount of IFN-γ produced by these cells was significantly decreased [126].

In addition to studies of cancer patients with solid tumors, Fujii *et al.* [33] have indicated that patients with myelodysplastic syndromes have a severe functional deficiency in NKT cells, but not NK cells or CD4$^+$ and CD8$^+$ T cells. Interestingly, freshly isolated from patients with progressive multiple myeloma and stimulated with α-GalCer NKT cells have a decreased capacity to produce IFN-γ as compared with functionally non-altered NKT cells of patients with non-progressive multiple myeloma or premalignant gammopathy [25]. Based on these findings, Seino *et al.* [104] have hypothesized that presentation of tumor-derived ligands by myeloma cells might cause NKT cell dysfunction *in vivo* and have recommended to restore the function of NKT cells in some cancer patients before NKT cell-mediated immunotherapy prescription. In this regard, it has been shown that preliminary application of granulocyte colony-stimulating factor (G-CSF) can partly restore the repressed NKT cell function in cancer patients [35] and probably improve subsequent NKT cell-based immunotherapy. Interestingly, *in vitro* activated DCs also can be successfully used to stimulate IFN-γ productivity of NKT cells [25, 143]. These reports, together with animal studies showing that a-GalCer-loaded DCs induce a prolonged IFN-γ-producing NKT-cell response *in vivo* [34], suggest a potential for DCs to renew activity of suppressed NKT cells in cancer patients. More detailed we discuss it below.

Furthermore, these reports indicate that NKT cells in cancer patients have some negative numerical and functional alterations, although it has not been elucidated whether Treg or other immunosuppressive mechanisms were involved in the clinical cases. Terabe *et al.* [136] have indicate that some observations of NKT cells from cancer patients are consistent with their observations in mouse tumor models, especially regarding to biased Th2 and deficient Th1 immune responses. Therefore, beneficial clinical results probably can be expected from immunotherapeutic strategies directed at expansion and activation of Th1-polarized NKT in cancer patients.

NKT Cell Ligands as Potential Candidates for Cancer Immunotherapy

A number of ligands that can be presented by CD1d-molecule to NKT cells have been identified (reviewed in [13]). α-GalCer is a glycosphingolipid that was originally derived from the marine sponge *Agelas mauritianus* [86] during a screen for novel antitumor agents. In 1997, Taniguchi and colleagues found that α-GalCer could stimulate murine NKT cells to rapidly produce both Th1 and Th2 cytokines in CD1d-dependent manner [51]. This study opened a floodgate of reports showing that α-GalCer could be effective as an antitumor immunopotentiator in a wide range of tumor model systems (reviewed in [14]). Now it is clear that α-GalCer binds CD1d-molecules from mouse and human and stimulates NKT cells from either species [12, 101, 121].

α-GalCer is not a physiological ligand of NKT cells because mammals and many other organisms are unable to synthesize α-anomeric glycosphingolipids. Nevertheless, α-GalCer might mimic "self" Ags that are recognized by NKT cells [150]. Although, accumulating evidence indicates that α-GalCer-activated NKT cells can kill a wide range of tumor target cells *in vitro* using cell-mediated perforin-dependent direct cytotoxicity [52], the predominant effector cells *in vivo* are the other immune system cells. Administration of α-GalCer to mice activates NKT and rapidly induces secretion of various cytokines, including IFN-γ and IL-4 [112]. Activation of NKT with α-GalCer also leads to upregulation of CD40L on NKT cells, leading to the activation of IL-12 production by DCs [55, 137]. α-GalCer-activated NKT cells also induce the maturation of DCs, which contributes to the upregulation of Th1 responses [31, 35]. IL-12 activates NKT cells to produce IFN-γ which in turn stimulates NK cells and $CD8^+$ cell-mediated antitumor cytotoxicity [46, 83, 117]. This is the cascade of events that is eventually believed to result in the development of a Th1-biased proinflammatory antitumor immune response that is pivotal for the antimetastatic activity of a-GalCer by promoting the generation of a long-lasting antitumor effector cell populations and by inhibiting tumor angiogenesis [38, 39, 47]. Indeed, emerging evidence indicates that IFN-γ and IL-12-activated NK cells play a critical role in the antimetastatic effects of α-GalCer [46, 114]. Thus, stimulation of NKT cells with α-GalCer results in activation as well as innate (NK cells and DCs) and adaptive (cytotoxic T cells and B cells) arms of immunity (reviewed in [37, 50]).

Recently, a series of structure-based synthetic α-GalCer analogues were developed and some of them were shown to be more effective than α-GalCer in cancer growth inhibition and selectively induced increased release of IL-4 and IFN-γ by human NKT cells *in vitro* [17]. These findings indicate that α-GalCer analogues can be designed to favor Th1-biased immunity, with greater anticancer efficacy and other immunomodulatory activities than natural α-GalCer.

There has also been some success in identification of foreign microbial glycolipid ligands of CD1d that can activate NKT cells (reviewed in [114]). Moreover, a strong candidate for a physiologically relevant natural ligand of NKT cells has been identified. This is a lysosomal glycolipid, isoglobotrihexosylceramide (iGb3), which has the ability to activate most human or mouse NKT cells *in vitro*. The absence of iGb3 in mice lacking the enzyme β-hexoaminidase B results in a dramatic loss of NKT cell number suggesting that this glycolipid may be a major selecting ligand for NKT cells *in vivo* [154].

NKT Cell-Based Cancer Immunotherapy Approaches and Results of First Clinical Trials

The full understanding of the immunobiology of cancer immunosurveillance will hopefully stimulate development of more effective immunotherapeutic approaches to control and/or eliminate human cancers [8]. Increasing evidence suggests that NKT cells occupy a unique intermediary position between the innate and acquired immune systems; therefore, the manipulation of NKT cells is expected to contribute to the development of therapeutic strategies for successful treatment of various human diseases including cancer [142, 151].

Antitumor effects of α-GalCer were observed in a variety of tumor metastasis models of liver, lung, and lymph nodes, including colon carcinoma, T-cell lymphoma, sarcoma, melanoma, and lung carcinoma, suggesting broad clinical applicability (reviewed in [142]). Clinical evaluation of the antitumor activity of iNKT cells was initiated in the Netherlands. A phase I study of direct intravenous injection of α-GalCer was carried out in patients with solid tumors [36]. Although no clinical responses were observed, in this study, no dose-limiting toxicity was observed, and the injection of α-GalCer over a wide range of doses was well tolerated in cancer patients. Also, this study indicated the importance of a sizeable pool of NKT cells in patients treated with α-GalCer, because an increase in serum cytokine levels (TNF-α and GM-CSF) only occurred in the smaller subset of patients with relatively high NKT cell frequency. Most recently, preliminary results of a phase I study of *in vitro* expanded autologous NKT cells in patients with advanced and recurrent non-small cell lung cancer were published [81]. Unfortunately, this strategy of NKT cell-based immunotherapy also did not result in clinical responses but it was well tolerated too.

Experimental evidence indicated that the antitumor activity of α-GalCer could be enhanced when α-GalCer was loaded onto DCs [1, 34, 110]. α-GalCer-pulsed DCs effectively activated NKT cells *in vivo*, resulting in the inhibition of tumor metastasis [138]. Based on the *in vivo* effects of α-GalCer-pulsed DCs, this immunotherapeutic approach has been evaluated in several clinical studies which showed that immunotherapy with α-GalCer-pulsed DCs seemed more potent and induced inflammatory tumor responses, tumor necrosis, and decreases in tumor markers. Also, in some patients, an expansion of NKT cells over several weeks to a few months and an increase in adaptive T-cell immunity were observed [16, 48, 87]. Again, immunologic responses were most prominent in patients with relatively high numbers of circulating NKT cells [48] and *in vivo* NKT cell responses were strongest using mature DCs [16]. Meantime Ishikawa *et al.* [49] proved the safety of α-GalCer-loaded DCs immunotherapy during a phase I study of treatment of patients with non-small cell lung cancer.

Van der Vliet *et al.* [141] have made examples of other approaches of NKT cell application, including the combination of NKT cell transfer followed by a-GalCer treatment (ideally in the context of DCs that can simultaneously be loaded with tumor-associated Ags), NKT therapy after chemotherapy pretreatment as this seems to sensitize tumor cells to NKT cell-mediated cytotoxicity [70], and direct modulation of CD1d on APC (e.g., by anti-CD1d monoclonal antibodies) as this has been shown to trigger the production of the Th1-promoting cytokine IL-12 [142]. Whereas Treg cells are frequently expanded in cancer patients, NKT cell-based immunotherapy therapy might need to be combined with CD4$^+$CD25$^+$ T-cell depletion (e.g., by denileukin diftitox) [23]. The recent findings that NKT cells are relatively resistant to apoptosis can be helpful for the rational design of NKT cell-based therapy protocols in the complex treatment of cancer patients.

NKT and Dendritic Cells are the Keys to Direct Antitumor Immune Response

NKT cells (type I) can be extremely useful to development the cell-based antitumor immunotherapy due to capacity to modulate DC function; that is why in presented part we will talk mostly about type I NKT cells. Just as myeloid-derived DCs mostly are applied for constructing of antitumor DC-based vaccines [139, 153] so the term "DCs" in this part of article we will use referring to myeloid-derived DCs. Interestingly, both DCs and NKT cells can polarize Th1 or Th2 immune response depending on cytokine environment and co-stimulation events [2, 104, 143] (Fig. 2) and that is the reason why these cells are considered as key cells of immune system [97, 104].

As was mentioned above, α-GalCer is a glycolipid Ag which activates immature NKT cells through restricted TCR. Kitamura *et al.* [55] for the first time showed that α-GalCer selectively activates NKT cells to produce INF-γ. This effect was dependent on IL-12 producing by DCs and CD40-CD40L interaction. Moreover, in response on α-GalCer NKT cells expressed markedly higher number of IL-12 receptor. Obviously, such interaction should lead to Th1 immune cell activation.

NKT cells interact with CD1d on DCs due to features of TCR [15, 26, 154]. CD1d is MHC-like molecule, which broadly surveys the early endocytic system for sampling lipid Ags [15] and is expressed on a wide spectrum of cells with different histogenesis [26]. A lot of literature data demonstrated, that DCs loaded with α-GalCer- activate NKT cells to produce Th1 cytokines and enhance cytolitic activity [65, 82, 130]. Stimulated with α-GalCer along or with α-GalCer-loaded DCs, NKT cells develop cytotoxic phenotype and produce a large amount of TNF-α and INF-γ – cytokines, which considered being essential for maturation and stimulation of naive DCs, and IL-12, which activate Th1 lymphocytes and NK cells. Matured DCs capture tumor Ag, direct to lymph nodes and spleen and present it to T-lymphocytes to initiate antitumor immune response [45]. Under maturation DCs produce high level of IL-12 and express a variety of co-stimulation molecules, particularly CD40, which, in turn, stimulate both NKT cells and cytotoxic T lymphocytes.

Normally, mentioned events lead to maturation and Ag capture by DCs, which migrate to lymph nodes and spleen, polarize CD4$^+$ cells to Th1 and direct naïve CD8$^+$ cells to cytotoxic T lymphocytes [32, 148] (Fig. 2). However, during tumor growth there are a number of mechanisms by which tumor cells can avoid detection and destruction by the immune system. These include down-regulation of HLA class I expression on the surface of tumor cells; down-regulation of tumor Ag processing and expression; lack of costimulatory molecules on tumors cells; release of immunosuppressive factors and other mechanisms [28, 99], even the site of the tumor origin has been described as factor that determines immunosuppressive effect [56, 106]. Some of the tumor-derived factors, such as vascular endothelial growth factor (VEGF) [62, 89], TGF-β [149], cyclooxygenase-2 [109], IL-10 [76], and others are able to down-regulate the generation and maturation of DCs and NKT cells [25, 40, 109], define character of DC-NKT cell interaction, and direct immune response on Th2 type [92]. Moreover, the number of IFN-γ-secreting Valpha24$^+$Vbeta11$^+$ NKT cell is decreased in cancer patients [77]. And one more important features of tumors appears to designate polarizing of NKT cells – tumors' microenvironment often contains Treg cells which also suppress all subsets of Vα24$^+$ NKT cells in both proliferation and cytokine production, both

Th1 and Th2 [3, 22]. This suppression is mediated by cell-to-cell contact but not by a humoral factor or the inhibition of APC. Moreover, the cytotoxic activity of Vα24⁺ NKT cells against some tumor cell lines is suppressed by CD4⁺CD25⁺ T cells [3], IL-13 producing NKT [95, 100] and myeloid suppressor cells [113]. So, the question is whether it is possible to activate NKT cells to produce Th1-determined factors and to interplay with DCs and T cells in cancer patients.

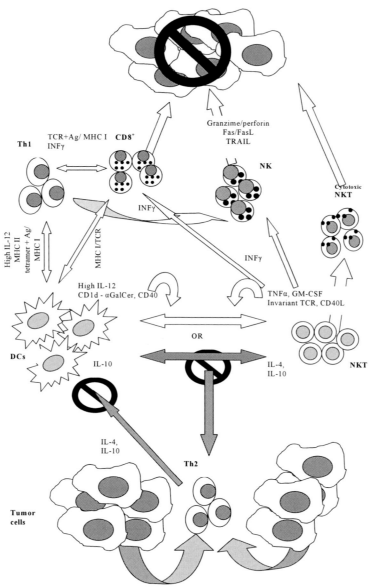

Figure 2. Antitumor effects provided by interplaying of DCs and NKT cells. DCs and NKT cells cooperate to increase Th1 immune response instead Th2, which lead to forming antitumor cytotoxic T cells, activation of NK and NKT cell cytolytic functions, and, in turn, destroying tumor cells.

Such inhibitory effects can be abrogated by using G-CSF or combination G-CSF and Flt-3L chimeric cytokine, progeninpoetin-1, when both NKT cell and DC function seems to be enhanced, as was demonstrated in P815 mastocytoma, non-Hodgkin's lymphoma and multiple myeloma [79, 147]. Another cytokine which is able to enhance NKT cell and DC activity to Th1 immune response and avoid surgically induced immune suppression in colorectal cancer patients appears INF-α [92]. Alternatively, it can be used so well-known Th1 stimulators as CpG DNA [90, 111], which action is double-directed towards NKT cells: it activates DCs to produce IL-12 and directly activates NKT cells to produce INF-γ [124].

To abrogate suppressing effects on NKT cells can be used α-GalCer, however in spite of the direct stimulatory effect of α-GalCer on NKT cells in tumor-bearing animals and *in vitro* on tumor cell cultures [53, 93, 96, 154], its effect is insufficient and may lead even to enhanced production of Th2 cytokines by NKT cells [92]. Nevertheless, dramatically promising has proved administration of α-GalCer-loaded or tumor specific Ag-loaded DCs to induce proliferation and Th1 cytokine production of NKT cells in tumor-bearing animals or cancer patients. Thus, Toura *et al.* [139] and Marten *et al.* [66] more then 10 years ago, and more recently Liu *et al.* [64] and van den Broeke *et al.* [140], showed that idiotype or α-GalCer-loaded DCs, or DCs activated by intravenous injection of irradiated tumor cells, effectively induced potent specific activation of NKT cells *in vitro* and *in vivo*. Due to this activation specific resistance both to myeloma and lymphoma was generated [64] and the inhibition of experimental melanoma and Lewis lung carcinoma [138] and CT26 colon carcinoma and as well Lewis lung carcinoma [140] metastasis was observed. Chang *et al.* [16] approved increase in serum level of IL-12 and INF-γ, and the number of specific cytotoxic T cells in response on administration of α-GalCer-loaded DCs. Nieda *et al.* [87] demonstrated that implantation of α-GalCer-loaded DCs stimulated both innate and adoptive antitumor effect in patients with breast, colon, liver cancer, melanoma, peritoneal adenocarcinoma, renal cell carcinoma, prostate carcinoma and lung carcinoma. Tatsumi *et al.* [130] and Nagaraj *et al.* [82] showed *in vivo* antitumor effect of α-GalCer-loaded DCs as regards CMS4 liver cancer and Panc02 pancreatic cancer correspondingly. Similar effects were demonstrated by Smyth *et al.* [120], who proved that combination of α-GalCer-loaded DCs and IL-12 is much more effective suppress B16F10 metastases in mice than α-GalCer alone or α-GalCer with IL-12. Van der Vliet *et al.* [143] used another cytokine, IL-15, to improve Th1 stimulatory effect of α-GalCer-loaded IL-12 producing DCs in cancer patients (ovarian cancer, non-small cell lung cancer, metastatic colon cancer, laryngeal cancer, metastatic melanoma, metastatic breast cancer, and metastatic renal cancer).

From all mentioned above we can see, that the most perspective ways for cancer immunotherapy are those which include cooperation between NKT cells and DCs. It promises safely, directed against tumor and tumor metastasis, and provides immune memory (in case of DCs).

Conclusion

The insight into the biology of NKT cells in general has expanded rapidly over recent years. Increasing evidence suggests that NKT cells occupy a unique intermediary position between innate and adaptive immune systems. Although NKT cells potentially can act as cytotoxic effector cells, it is likely that regulatory function reflects their true physiologic role.

This is consistent with their small number, the distinctive feature to produce different cytokines, and their close relationships with DCs, Treg and other cell types, and finally the remarkable functional diversity of NKT cells in various immune responses.

Significant advances have been recently made in understanding the functions of NKT cells for tumor immunosurveillance. Exciting enigma of controversial behavior of NKT cells in tumor immunosurveillance has been solved when different types of CD1d-restricted NKT cells contributing to different ways of antitumor immune response were discovered. Although important progress has been made regarding the specificity and effector functions of NKT cells, there are still many questions remaining to be explored. One of them is the possible relationship among the various immunoregulatory T cells: NKT, Treg and γδT cells. Future studies should also aim to elucidate the antitumor immune functions of distinct NKT cell subsets and subtypes because it can be extremely important to future development of successful approaches to NKT cell-based immunotherapy.

Understanding regulatory mechanisms of antitumor immunity is a desired key to development of successful cancer immunotherapy approaches. As NKT cells are highly potent immunoregulatory cells and are therefore very attractive targets for cancer immunotherapy. The use of NKT cell ligands, such as a-GalCer, has shown substantial beneficial effects in various tumor animal models. A phase I trials of α-Gal-Cer treatment or transfer of *in vitro* expanded and activated NKT cells in patients with various forms of cancer were carried out. Although positive clinical responses have not been observed, the absence of dose limiting toxicity and increased NKT cell activation and cytokine production were evident, suggesting that the potential for antitumor efficacy of such NKT cell-based immunotherapeutic approaches will be found in more refined future studies. Moreover, recently evidence suggested that, in combination with DC-based vaccination, these NKT cell-based immunotherapeutic approaches acquire the potential to substantially stimulate the development of conventional long-lasting antitumor immune responses. In the nearest future, reasonable manipulations with NKT cells can open up additional avenues for development of successful strategy of cancer patient's treatment.

References

[1]   Akutsu, Y; Nakayama, T; Harada, M; Kawano, T; Motohashi, S; Shimizu, E; Ito, T; Kamada, N; Saito, T; Matsubara, H; Miyazawa, Y; Ochiai, T; Taniguchi, M. Expansion of lung Valpha14 NKT cells by administration of alpha-galactosylceramide-pulsed dendritic cells. *Jpn J Cancer Res,* 2002 93, 397–403.

[2]   Amati, L; Pepe, M; Passeri, ME; Mastronardi, ML; Jirillo, E; Covelli, V. Toll-like receptor signaling mechanisms involved in dendritic cell activation: potential therapeutic control of T cell polarization. *Curr Pharm Des,* 2006 12, 4247–4254.

[3]   Azuma, T; Takahashi, T; Kunisato, A; Kitamura, T; Hirai, H; Human CD4$^+$CD25$^+$ regulatory T cells suppress NKT cell functions. *Cancer Res,* 2003 63, 4516–4520.

[4]   Behar, SM; Porceli, SA. CD1-restricted T cells in host defense to infectious diseases. *Curr Top Microbiol Immunol,* 2007 314, 215–250.

[5]   Bendelac, A; Killeen, N; Littman, DR; Schwartz, RH. A subset of CD4R thymocytes selected by MHC class I molecules. *Science,* 1994 263, 1774–1778.

[6]   Bendelac, A; Rivera, MN; Park, SH; Roark, JH. Mouse CD1-specific NK1 T cells: development, specificity, and function. *Annu Rev Immunol,* 1997 15, 535–562.

[7]   Benlagha, K; Kyin, T; Beavis, A; Teyton, L; Bendelac, A. A thymic precursor to the NKT cell lineage. *Science,* 2002 296, 553–555.

[8]   Bhardwaj, N. Harnessing the immune system to treat cancer. *J Clin Invest,* 2007 117, 1130–1136.

[9]   Biron, CA; Brossay, L. NK cells and NKT cells in innate defense against viral infections. *Curr Opin Immunol,* 2001 13, 458–464.

[10]   Brigl, M; Brenner, MB. CD1: antigen presentation and T cell function. *Annu Rev Immunol,* 2004 22, 817–890.

[11]   Brigl, M; Bry, L; Kent, SC; Gumperz, JE; Brenner, MB. Mechanism of CD1d-restricted natural killer T cell activation during microbial infection. *Nat Immunol,* 2003 4, 1230–1237.

[12]   Brossay, L; Chioda, M; Burdin, N; Koezuka, Y; Casorati, G; Dellabona, P; Kronenberg, M. CD1d-mediated recognition of an α-galactosylceramide by natural killer T cells is highly conserved through mammalian evolution. *J Exp Med,* 1998 188, 1521–1528.

[13]   Brutkiewicz, RR. CD1d ligands: the good, the bad, and the ugly. *J Immunol,* 2006 177, 769–775.

[14]   Brutkiewicz, RR; Sriram, V. Natural killer T (NKT) cells and their role in antitumor immunity. *Crit Rev Oncol Hematol,* 2002 41, 287–298.

[15]   Cao, X; Sugita, M; Van Der Wel, N; Lai, J; Rogers, RA; Peters, PJ; Brenner, MB. CD1 molecules efficiently present antigen in immature dendritic cells and traffic independently of MHC class II during dendritic cell maturation. *J Immunol,* 2002 169, 4770–4777.

[16]   Chang, DH; Osman, K; Connolly, J; Kukreja, A; Krasovsky, J; Pack, M; Hutchinson, A; Geller, M; Liu, N; Annable, R; Shay, J; Kirchhoff, K; Nishi, N; Ando, Y; Hayashi, K; Hassoun, H; Steinman, RM; Dhodapkar, MV. Sustained expansion of NKT cells and antigen-specific T cells after injection of a-galactosylceramide loaded mature dendritic cells in cancer patients. *J Exp Med,* 2005 201, 1503–1517.

[17]   Chang, YJ; Huang, JR; Tsai, YC; Hung, JT; Wu, D; Fujio, M; Wong, CH; Yu, AL. Potent immune-modulating and anticancer effects of NKT cell stimulatory glycolipids. *Proc Natl Acad Sci USA,* 2007 104, 10299–10304.

[18]   Chiu, YH; Park, SH; Benlagha, K; Forestier, C; Jayawardena-Wolf, J; Savage, PB; Teyton, L; Bendelac, A. Multiple defects of antigen presentation and T cell development in mice expressing tail-truncated CD1d. *Nat Immunol,* 2002 3, 55–60.

[19]   Coppieters, K; Dewint, P; van Beneden, K; Jacques, P; Seenws, S; Verbrugsen, G; Deforce, D; Elewant, D. NKT cells: manipulable managers of joint inflammation. *Rheumatology (Oxford),* 2007 46, 565–571.

[20]   Crowe, NY; Coquet, JM; Berzins, SP; Kyparissoudis, K; Keating, R; Pellicci, DG; Hayakawa, Y; Godfrey, DI; Smyth, MJ. Differential antitumor immunity mediated by NKT cell subsets *in vivo. J Exp Med* 2005 202, 1279–1288.

[21]   Cui, J; Shin, T; Kawano, T; Sato, H; Kondo, E; Toura, I; Kaneko, Y; Koseki, H; Kanno, M; Taniguchi, M. Requirement for Valpha14 NKT cells in IL-12-mediated rejection of tumors. *Science,* 1997 278, 1623–1626.

[22] Danese, S; Rutella, S. The Janus face of CD4+CD25+ regulatory T cells in cancer and autoimmunity. *Curr Med Chem*, 2007 14, 649–666.

[23] Dannull, J; Su, Z; Rizzieri, D; Yang, BK; Coleman, D; Yancey, D; Zhang, A; Dahm, P; Chao, N; Gilboa, E; Vieweg, J. Enhancement of vaccine-mediated antitumor immunity in cancer patients after depletion of regulatory T cells. *J Clin Invest*, 2005 115, 3623–3633.

[24] Dellabona, P; Padovan, E; Casorati, G; Brockhaus, M; Lanzavecchia, A. An invariant Vα24-JαQ/Vβ11 T cell receptor is expressed in all individuals by clonally expanded CD4⁰CD8⁰ T cells. *J Exp Med*, 1994 180, 1171–1176.

[25] Dhodapkar, MV; Geller, MD; Chang, DH; Shimizu, K; Fujii, S; Dhodapkar, KM; Krasovsky, J. A reversible defect in natural killer T cell function characterizes the progression of premalignant to malignant multiple myeloma. *J Exp Med*, 2003 197, 1667–1676.

[26] Dougan, SK; Kaser, A, Blumberg RS. CD1 expression on antigen-presenting cells. *Curr Top Microbiol Immunol*, 2007 314, 113–141.

[27] Dunn, GP; Old, LJ; Schreiber1, RD. The immunobiology of cancer immunosurveillance and immunoediting. *Immunity*, 2004 21, 137–148.

[28] Esche, C; Lokshin, A; Shurin, GV; Gastman, BR; Rabinowich, H; Watkins, SC; Lotze, MT; Shurin, MR. Tumor's other immune targets: dendritic cells. *J Leukoc Biol*, 1999 66, 336–344.

[29] Exley, M; Garcia, J; Balk, SP; Porcelli, S. Requirements for CD1d recognition by human invariant Vβ24+ CD4⁰CD8⁰ T cells. *J Exp Med*, 1997 186, 109–120.

[30] Fowlkes, BJ; Kruisbeek, AM; Ton-That, H; Weston, MA; Coligan, JE; Schwartz, RH; Pardoll, DM. A novel population of T-cell receptor ab-bearing thymocytes which predominantly expresses a single Vb gene family. *Nature*, 1987 329, 251–254.

[31] Fujii, S; Liu, K; Smith, C; Bonito, AJ; Steinman, RM. The linkage of innate to adaptive immunity via maturing dendritic cells in vivo requires CD40 ligation in addition to antigen presentation and CD80/86 costimulation. *J Exp Med*, 2004 199, 1607–1618.

[32] Fujii, S; Shimizu, K; Hemmi, H; Fukui, M; Bonito, AJ; Chen, G; Franck, RW; Tsuji, M; Steinman, RM. Glycolipid alpha-C-galactosylceramide is a distinct inducer of dendritic cell function during innate and adaptive immune responses of mice. *Proc Natl Acad Sci USA*, 2006 103, 11252–11257.

[33] Fujii, S; Shimizu, K; Klimek, V; Geller, MD; Nimer, SD; Dhodapkar, MV. Severe and selective deficiency of interferon-gamma-producing invariant natural killer T cells in patients with myelodysplastic syndromes. *Br J Haematol*, 2003 122, 617–622.

[34] Fujii, S; Shimizu, K; Kronenberg, M; Steinman, RM. Prolonged IFN-gamma-producing NKT response induced with alpha-galactosylceramide-loaded DCs. *Nat Immunol*, 2002 3, 867–874.

[35] Fujii, S; Shimizu, K; Smith, C; Bonifaz, L; Steinman, RM. Activation of natural killer T cells by alpha-galactosylceramide rapidly induces the full maturation of dendritic cells in vivo and thereby acts as an adjuvant for combined CD4 and CD8 T cell immunity to a coadministered protein. *J Exp Med*, 2003 198, 267–279.

[36] Giaccone, G; Punt, CJ; Ando, Y; Ruijter, R; Nishi, N; Peters, M; von Blomberg, BM; Scheper, RJ; van der Vliet, HJ; van den Eertwegh, AJ; Roelvink, M; Beijnen, J; Zwierzina, H; Pinedo, HM. A phase I study of the natural killer T-cell ligand alpha-

galactosylceramide (KRN7000) in patients with solid tumors. *Clin Cancer Res*, 2002 8, 3702–3709.

[37]   Godfrey, DI; Hammond, KJ; Poulton, LD; Smyth, MJ; Baxter, AG. NKT cells: facts, functions and fallacies. *Immunol Today*, 2000 21, 573–583.

[38]   Godfrey, DI; Kronenberg, M. Going both ways: immune regulation via CD1d-dependent NKT cells. *J Clin Invest*, 2004 114, 1379–1388.

[39]   Godfrey, DI; MacDonald, HR; Kronenberg, M; Smyth, MJ; Van Kaer, L. NKT cells: what's in a name? *Nat Rev Immunol*, 2004 4, 231–237.

[40]   Gumperz, JE. CD1d-restricted "NKT" cells and myeloid IL-12 production: an immunological crossroads leading to promotion or suppression of effective anti-tumor immune responses? *J Leukoc Biol*, 2004 76, 307–313.

[41]   Gumperz, JE; Brenner, MB. CD1-specific T cells in microbial immunity. *Curr Opin Immunol*, 2001 13, 471–478.

[42]   Gumperz, JE; Miyake, S; Yamamura, T; Brenner, MB. Functionally distinct subsets of CD1d-restricted natural killer T cells revealed by CD1d tetramer staining. *J Exp Med*, 2002 195, 625–636.

[43]   Gumperz, JE; Roy, C; Makowska, A; Lum, D; Sugita, M; Podrebarac, T; Koezuka, Y; Porcelli, SA; Cardell, S; Brenner, MB; Behar, SM. Murine CD1d-restricted T cell recognition of cellular lipids. *Immunity*, 2000 12, 211–221.

[44]   Hammond, KJL; Kronenberg, M. Natural killer T cells: natural or unnatural regulators of autoimmunity? *Curr Opin Immunol*, 2003 15, 683–689.

[45]   Hart, DNJ. Dendritic cells, unique leucocyte populations which control the primary immune response. *Blood*, 1997 90, 3245–3287.

[46]   Hayakawa, Y; Takeda, K; Yagita, H; Kakuta, S; Iwakura, Y; Van Kaer, L; Saiki, I; Okumura, K. Critical contribution of IFN gamma and NK cells, but not perforin-mediated cytotoxicity, to antimetastatic effect of alpha-galactosylceramide. *Eur J Immunol*, 2001 31, 1720–1727.

[47]   Hayakawa, Y; Takeda, K; Yagita, H; Smyth, MJ; Van Kaer, L; Okumura, K; Saiki, I. IFN-gamma-mediated inhibition of tumor angiogenesis by natural killer T-cell ligand, alpha-galactosylceramide. *Blood*, 2002 100, 1728–1733.

[48]   Ishikawa, A; Motohashi, S; Ishikawa, E; Fuchida, H; Higashino, K; Otsuji, M; Iizasa, T; Nakayama, T; Taniguchi, M; Fujisawa, T. A phase I study of alpha-galactosylceramide (KRN7000)-pulsed dendritic cells in patients with advanced and recurrent non-small cell lung cancer. *Clin Cancer Res*, 2005 11,1910–1917.

[49]   Ishikawa, E; Motohashi, S; Ishikawa, A; Ito, T; Uchida, T; Kaneko, T; Tanaka, Y; Horiguchi, S; Okamoto, Y; Fujisawa, T; Tsuboi, K; Taniguchi, M; Matsumura, A; Nakayama, T. Dendritic cell maturation by CD11c− T cells and Valpha24+ natural killer T-cell activation by alphagalactosylceramide. *Int J Cancer*, 2005 117, 265–273.

[50]   Joyce, S. CD1d and natural T cells: how their properties jump-start the immune system. *Cell Mol Life Sci*, 2001 58, 442–469.

[51]   Kawano, T; Cui, J; Koezuka, Y; Toura, I; Kaneko, Y; Motoki, K; Ueno, H; Nakagawa, R; Sato, H; Kondo, E; Koseki, H; Taniguchi, M. CD1d-restricted and TCR-mediated activation of Vα14 NKT cells by glycosylceramides. *Science*, 1997 278, 1626–1629.

[52]   Kawano, T; Cui, J; Koezuka, Y; Toura, I; Kaneko, Y; Sato, H; Kondo, E; Harada, M; Koseki, H; Nakayama, T; Tanaka, Y; Taniguchi, M. Natural killer-like nonspecific

tumor cell lysis mediated by specific ligand-activated Valpha14 NKT cells. *Proc Natl Acad Sci USA*, 1998 95, 5690–5693.

[53]   Kawano, T; Nakayama, T; Kamada, N; Kaneko, Y; Harada, M; Ogura, N; Akutsu, Y; Motohashi, S; Iizasa, T; Endo, H; Fujisawa, T; Shinkai, H; Taniguchi, M. Antitumor cytotoxicity mediated by ligand-activated human V alpha24 NKT cells. *Cancer Res*, 1999 59, 5102–5105.

[54]   Kenna, T; Mason, LG; Porcelli, SA; Koezuka, Y; Hegarty, JE; O'Farrelly, C; Doherty, DG. NKT cells from normal and tumorbearing human livers are phenotypically and functionally distinct from murine NKT cells. *J Immunol*, 2003 171, 1775–1779.

[55]   Kitamura, H; Iwakabe, K; Yahata, T; Nishimura, S; Ohta, A; Ohmi, Y; Sato, M; Takeda, K; Okumura, K; Van Kaer, L; Kawano, T; Taniguchi, M; Nishimura, T. The natural killer T (NKT) cell ligand alpha-galactosylceramide demonstrates its immunopotentiating effect by inducing interleukin (IL)-12 production by dendritic cells and IL-12 receptor expression on NKT cells. *J Exp Med*, 1999 189, 1121–1128.

[56]   Ko, YH; Cho, EY; Kim, JE; Lee, SS; Huh, JR; Chang, HK; Yang, WI; Kim, CW; Kim, SW; Ree, HJ. NK and NK-like T-cell lymphoma in extranasal sites: a comparative clinicopathological study according to site and EBV status. *Histopathology*, 2004 44, 480–489.

[57]   Koseki, H; Imai, K; Ichikawa, T; Hayata, I; Taniguchi, M. Predominant use of a particular alpha-chain in suppressor T cell hybridomas specific for keyhole limpet hemocyanin. *Int Immunol*, 1989 1, 557–564.

[58]   Koseki, H; Asano, H; Inaba, T; Miyashita, N; Moriwaki, K; Lindahl, KF; Mizutani, Y; Imai, K; Taniguchi, M. Dominant expression of a distinctive V14$^+$ T-cell antigen receptor a chain in mice. *Proc Natl Acad Sci USA*, 1991 88, 7518–7522.

[59]   Kronenberg, M. Toward an understanding of NKT cell biology: progress and paradoxes. *Annu Rev Immunol*, 2005 26, 877–900.

[60]   Kronenberg, M; Gapin, L. The unconventional lifestyle of NKT cells. *Nat Rev Immunol*, 2002 2, 557–568.

[61]   Kronenberg, M; Rudensky, A. Regulation of immunity by self-reactive T cells. *Nature*, 2005 435, 598–604.

[62]   Kusmartsev, S; Gabrilovich, DI. Immature myeloid cells and cancer-associated immune suppression. *Cancer Immunol Immunother*, 2002 51, 293–298.

[63]   La Cava, A; Van Kaer, L; Fu-Dong-Shi. CD4$^+$CD25$^+$ Tregs and NKT cells: regulators regulating regulators. *TRENDS Immunol*, 2006 27, 322–327.

[64]   Liu, K; Idoyaga, J; Charalambous, A; Fujii, S; Bonito, A; Mordoh, J; Wainstok, R; Bai, XF; Liu, Y; Steinman, RM. Innate NKT lymphocytes confer superior adaptive immunity via tumor-capturing dendritic cells. *J Exp Med*, 2005 202, 1507–1516.

[65]   Lu, X; Song, L; Metelitsa, LS; Bittman, R. Synthesis and evaluation of an alpha-C-galactosylceramide analogue that induces Th1-biased responses in human natural killer T cells. *Chembiochem*, 2006 7, 1750–1756.

[66]   Marten, A; Renoth, S; von Lilienfeld-Toal, M; Buttgereit, P; Schakowski, F; Glasmacher, A; Sauerbruch, T; Schmidt-Wolf, IG. Enhanced lytic activity of cytokine-induced killer cells against multiple myeloma cells after co-culture with idiotype-pulsed dendritic cells. *Haematologica*, 2001 86, 1029–1037.

[67] Matsuda, JL; Gapin, L. Does the developmental status of Va14i NKT cells play a role in disease? *Int Rev Immunol*, 2007 26, 5–29.

[68] Matsuda, JL; Gapin, L; Baron, JL; Sidobre, S; Stetson, DB; Mohrs, M; Locksley, RM; Kronenberg, M. Mouse V alpha 14i natural killer T cells are resistant to cytokine polarization in vivo. *Proc Natl Acad Sci USA*, 2003 100, 8395–8400.

[69] Matsui, S; Ahlers, JD; Vortmeyer, AO; Terabe, M; Tsukui, T; Carbone, DP; Liotta, LA; Berzofsky, J. A model for CD8$^+$ CTL tumor immunosurveillance and regulation of tumor escape by CD4 T cells through an effect on quality of CTL. *J Immunol*, 1999 163, 184–193.

[70] Mattarollo, SR; Kenna, T; Nieda, M; Nicol, AJ. Chemotherapy pretreatment sensitizes solid tumor-derived cell lines to Va24$^+$ NKT cell-mediated cytotoxicity. *Int J Cancer*, 2006 119, 1630–1637.

[71] Mendiratta, SK; Martin, WD; Hong, S; Boeseanu, A; Joyce, S; van Kaer, L. CD1d1 mutant mice are deficient in natural T cells that promptly produce IL-4. *Immunity*, 1997 6, 469–477.

[72] Metelitsa, LS; Naidenko, OV; Kant, A; Wu, HW; Loza, MJ; Perussia, B; Kronenberg, M; Seeger, RC. Human NKT cells mediate antitumor cytotoxicity directly by recognizing target cell CD1d with bound ligand or indirectly by producing IL-2 to activate NK cells. *J Immunol*, 2001 167, 3114–3122.

[73] Metelitsa, LS; Wu, HW; Wang, H; Yang, Y; Warsi, Z; Asgharzadeh, S; Groshen, S; Wilson, SB; Seeger, RC. Natural killer T cells infiltrate neuroblastomas expressing the chemokine CCL2. *J Exp Med*, 2004 199, 1213–1221.

[74] Meyer, EH; DeKruyff, RH; Umetsu, DT. iNKT cells in allergic disease. *Curr Top Microbiol Immunol*, 2007 314, 269–291.

[75] Mocellin, S; Marincola, FM; Young, HA. Interleukin-10 and the immune response against cancer: a counterpoint. *J Leukoc Biol*, 2005 78, 1043–1051.

[76] Molano, A; Porcelli, SA. Invariant NKT cell regulation of autoimmunity. *Drug Discov Today*, 2006 3, 193–198.

[77] Molling, JW; Kolgen, W; van der Vliet, HJ; Boomsma, MF; Kruizenga, H; Smorenburg, CH; Molenkamp, BG; Langendijk, JA; Leemans, CR; von Blomberg, BM; Scheper, RJ; van den Eertwegh, AJ. Peripheral blood IFN-gamma-secreting Valpha24$^+$Vbeta11$^+$ NKT cell numbers are decreased in cancer patients independent of tumor type or tumor load. *Int J Cancer*, 2005 116, 87–93.

[78] Moodycliffe, AM; Nghiem, D; Clydesdale, G; Ullrich, SE. Immune suppression and skin cancer development: regulation by NKT cells. *Nat Immunol*, 2000 1, 521–525.

[79] Morris, ES; MacDonald, KP; Rowe, V; Banovic, T; Kuns, RD; Don, AL; Bofinger, HM; Burman, AC; Olver, SD; Kienzle, N; Porcelli, SA; Pellicci, DG; Godfrey, DI; Smyth, MJ; Hill, GR. NKT cell-dependent leukemia eradication following stem cell mobilization with potent G-CSF analogs. *J Clin Invest*, 2005 115, 3093–3103.

[80] Motohashi, S; Kobayashi, S; Ito, T; Magara, KK; Mikuni, O; Kamada, N; Iizasa, T; Nakayama, T; Fujisawa, T; Taniguchi, M. Preserved IFN-gamma production of circulating Valpha24 NKT cells in primary lung cancer patients. *Int J Cancer*, 2002 102, 159–165.

[81] Motohashi, S.;, Ishikawa, A; Ishikawa, E; Otsuji, M; Iizasa, T; Hanaoka, H; Shimizu, N; Horiguchi, S; Okamoto, Y; Fujii, S; Taniguchi, M; Fujisawa, T; Nakayama, T. A

phase I study of in vitro expanded natural killer T cells in patients with advanced and recurrent non-small cell lung cancer. *Clin Cancer Res*, 2006 12, 6079–6086.

[82]   Nagaraj, S; Ziske, C; Strehl, J; Messmer, D; Sauerbruch, T; Schmidt-Wolf, IG. Dendritic cells pulsed with alpha-galactosylceramide induce anti-tumor immunity against pancreatic cancer in vivo. *Int Immunol*, 2006 18, 1279–1283.

[83]   Nakagawa, R; Nagafune, I; Tazunoki, Y; Ehara, H; Tomura, H; Iijima, R; Motoki, K; Kamishohara, M; Seki, S. Mechanisms of the antimetastatic effect in the liver and of the hepatocyte injury induced by alpha-galactosylceramide in mice. *J Immunol*, 2001 166, 6578–6584.

[84]   Nakamura, E; Kubota, H; Sato, M; Sugie, T; Yoshida, O; Minato, N. Involvement of NK1$^+$CD4$\Box$CD8$\Box$ αβ T cells and endogenous IL-4 in non-MHC-restricted rejection of embryonal carcinoma in genetically resistant mice. *J Immunol*, 1997 158, 5338–5348.

[85]   Natori, T; Koezuka, Y; Higa, T. Agelasphins, novel α-galactosylceramides from the marine sponge *Agelas mauritianus. Tetrahedron Lett*, 1993 34, 5591–5592.

[86]   Nieda, M; Nicol, A; Koezuka, Y; Kikuchi, A; Lapteva, N; Tanaka, Y; Tokunaga, K; Suzuki, K; Kayagaki, N; Yagita, H; Hirai, H; Juji, T. TRAIL expression by activated human CD4(+) Valpha24 NKT cells induces in vitro and in vivo apoptosis of human acute myeloid leukemia cells. *Blood*, 2001 97, 2067–2074.

[87]   Nieda, M; Okai, M; Tazbirkova, A; Lin, H; Yamaura, A; Ide, K; Abraham, R; Juji, T; Macfarlane, DJ; Nicol, AJ. Therapeutic activation of Valpha24$^+$Vbeta11$^+$ NKT cells in human subjects results in highly coordinated secondary activation of acquired and innate immunity. *Blood*, 2004 103, 383–389.

[88]   Nishikawa, H; Kato, T; Tanida, K; Hiasa, A; Tawara, I; Ikeda, H; Ikarashi, Y; Wakasugi, H; Kronenberg, M; Nakayama, T; Taniguchi, M; Kuribayashi, K; Old, LJ; Shiku, H. CD4$^+$CD25$^+$ T cells responding to serologically defined autoantigens suppress antitumor immune responses. *Proc Natl Acad Sci USA*, 2003 100, 10902–10906.

[89]   Ohm, JE; Carbone, DP. VEGF as a mediator of tumor-associated immunodeficiency. *Immunol Res*, 2001 23, 263–272.

[90]   Olishevsky, SV; Shlyakhovenko, VA; Kozak, VV; Yanish, YV. Immunostimulatory CpG DNA in cancer vaccinotherapy. *Exp Oncol* 2003 25, 83–92.

[91]   Oosterling, SJ; van der Bij, GJ; Mels, AK; Beelen, RH; Meijer, S; van Egmond, M; van Leeuwen, PA. Perioperative IFN-alpha to avoid surgically induced immune suppression in colorectal cancer patients. *Histol Histopathol*, 2006 21, 753–760.

[92]   Osada, T; Morse, MA; Lyerly, HK; Clay, TM. Ex vivo expanded human CD4$^+$ regulatory NKT cells suppress expansion of tumor antigen-specific CTLs. *Int Immunol*, 2005 17, 1143–1155.

[93]   Osada, T; Nagawa, H; Shibata, Y. Tumor-infiltrating effector cells of alpha-galactosylceramide-induced antitumor immunity in metastatic liver tumor. *J Immune Based Ther Vaccines*, 2004 2, 7.

[94]   Ostrand-Rosenberg, S; Clements, VK; Terabe, M; Park, JM; Berzofsky, J; Dissanayake, SK. Resistance to metastatic disease in Stat6-deficient mice requires hematopoietic and nonhematopoietic cells and is IFN-gamma dependent. *J Immunol*, 2002 169, 5796–5804.

[95]   Park, JM; Terabe, M; van den Broeke, LT; Donaldson, DD; Berzofsky, JA. Unmasking immunosurveillance against a syngeneic colon cancer by elimination of CD4⁺ NKT regulatory cells and IL-13. *Int J Cancer*, 2005 114, 80–87.

[96]   Park, SH; Kyin, T; Bendelac, A; Carnaud, C. The contribution of NKT cells, NK cells, and other gamma-chain-dependent non-T non-B cells to IL-12-mediated rejection of tumors. *J Immunol*, 2003 170,1197–1201.

[97]   Pearce, EJ; Kane, CM; Sun, J. Regulation of dendritic cell function by pathogen-derived molecules plays a key role in dictating the outcome of the adaptive immune response. *Chem Immunol Allergy*, 2006 90, 82–90.

[98]   Pellici, DG; Hammond, KJ; Uldrich, AP; Baxter, AG; Smyth, MJ; Godfrey, DI. A natural killer T (NKT) cell developmental pathway involving a thymus-dependent NK1.1⁻CD4⁺ CD1d-dependent precursor stage. *J. Exp. Med.*, 2002 195, 835–844.

[99]   Platsoucas, CD; Fincke, JE; Pappas, J; Jung, WJ; Heckel, M; Schwarting, R; Magira, E; Monos, D; Freedman, RS. Immune responses to human tumors: development of tumor vaccines. *Anticancer Res*, 2003 23, 1969–1996.

[100]   Renukaradhya, GJ; Sriram, V; Du, W; Gervay-Hague, J; Van Kaer, L; Brutkiewicz, RR. Inhibition of antitumor immunity by invariant natural killer T cells in a T-cell lymphoma model in vivo. *Int J Cancer*, 2006 118, 3045–3053.

[101]   Rogers, PR; Matsumoto, A; Naidenko, O; Kronenberg, M; Mikayama, T; Kato, S. Expansion of human Valpha24⁺ NKT cells by repeated stimulation with KRN7000. *J Immunol Methods*, 2004 285, 197–214.

[102]   Sakuishi, K; Oki, S; Araki, M; Porcelli, SA; Miyake, S; Yamamura, T; Invariant NKT cells biased for IL-5 production act as crucial regulators of inflammation. *J Immunol*, 2007 179, 3552–3562.

[103]   Seino, K; Harada, M; Taniguchi, M. NKT cells are relatively resistant to apoptosis. *TRENDS Immunol*, 2004 25, 219–221.

[104]   Seino, K; Motohashi, S; Fujisawa, T; Nakayama, T; Taniguchi, M. Natural killer T cell-mediated antitumor responses and their clinical applications. *Cancer Sci*, 2006 97, 807–812.

[105]   Seino, K; Taniguchi, M. Functionally distinct NKT cell subsets and subtypes. *J Exp Med*, 2005 202, 1623–1626.

[106]   Sevko, AL; Barysik, N; Perez, L; Shurin, MR; Gerein, V. Differences in dendritic cell activation and distribution after intravenous, intraperitoneal, and subcutaneous injection of lymphoma cells in mice. *Adv Exp Med Biol*, 2007 601, 257–264.

[107]   Shankaran, V; Ikeda, H; Bruce, AT; White, JM; Swanson, PE; Old, LJ; Schreiber, RD. IFN-γ and lymphocytes prevent primary tumour development and shape tumour immunogenicity. *Nature*, 2001 410 1107–1111.

[108]   Sharma, S; Stolina, M; Yang, SC; Baratelli, F; Lin, JF; Atianzar, K; Luo, J; Zhu, L; Lin, Y; Huang, M; Dohadwala, M; Batra, RK; Dubinett, SM. Tumor cyclooxygenase 2-dependent suppression of dendritic cell function. *Clin Cancer Res*, 2003 9, 961–968.

[109]   Shimizu, K; Hidaka, M; Kadowaki, N; Makita, N; Konishi, N; Fujimoto, K; Uchiyama, T; Kawano, F; Taniguchi, M; Fujii, S. Evaluation of the function of human invariant NKT cells from cancer patients using alpha-galactosylceramide-loaded murine dendritic cells. *J Immunol*, 2006 177, 3484–3492.

[110] Shin, T; Nakayama, T; Akutsu, Y; Motohashi, S; Shibata, Y; Harada, M; Kamada, N; Shimizu, C; Shimizu, E; Saito, T; Ochiai, T; Taniguchi, M. Inhibition of tumor metastasis by adoptive transfer of IL-12-activated Va14 NKT cells. *Int J Cancer*, 2001 91, 523–528.

[111] Shlyakhovenko, VA; Olishevsky, SV; Kozak, VV; Yanish, YV; Rybalko, SL. Anticancer and immunostimulatory effects of nucleoprotein fraction of *Bacillus subtilis* 7025 culture medium filtrate. *Exp Oncol* 2003 25, 119–23.

[112] Singh, N; Hong, S; Scherer, DC; Serizawa, I; Burdin, N; Kronenberg, M; Koezuka, Y; Van Kaer, L. Activation of NK T cells by CD1d and α-galactosylceramide directs conventional T cells to the acquisition of a Th2 phenotype. *J Immunol*, 1999 163, 2373–2377.

[113] Sinha, P; Clements, VK; Ostrand-Rosenberg, S. Interleukin-13-regulated M2 macrophages in combination with myeloid suppressor cells block immune surveillance against metastasis. *Cancer Res*, 2005 65, 11743–11751.

[114] Smyth, MJ; Cretney, E; Takeda, K; Wiltrout, RH; Sedger, LM; Kayagaki, N; Yagita, H; Okumura, K. Tumor necrosis factor-related apoptosis-inducing ligand (TRAIL) contributes to interferon gamma-dependent natural killer cell protection from tumor metastasis. *J Exp Med*, 2001 193, 661–670.

[115] Smyth, MJ; Crowe, NY; Godfrey, DI. NK cells and NKT cells collaborate in host protection from methylcholanthrene-induced fibrosarcoma. *Int Immunol*, 2001 13, 459–463.

[116] Smyth, MJ; Crowe, NY; Hayakawa, Y; Takeda, K; Yagita, H; Godfrey, DI. NKT cells – conductors of tumor immunity? *Curr Opin Immunol*, 2002 14, 165–171.

[117] Smyth, MJ; Crowe, NY; Pellicci, DG; Kyparissoudis, K; Kelly, JM; Takeda, K; Yagita, H; Godfrey, DI. Sequential production of interferon-gamma by NK1.1[+] T cells and natural killer cells is essential for the antimetastatic effect of alpha-galactosylceramide. *Blood*, 2002 99, 1259–1266.

[118] Smyth, MJ; Godfrey, DI. NKT cells and tumor immunity – a double-edged sword. *Nature Immunol*, 2000 1 6, 459–460.

[119] Smyth, MJ; Thia, KY; Street, SE; Cretney, E; Trapani, JA; Taniguchi, M; Kawano, T; Pelikan, SB; Crowe, NY; Godfrey, DI. Differential tumor surveillance by natural killer (NK) and NKT cells. *J Exp Med*, 2000 191, 661–668.

[120] Smyth, MJ; Wallace, ME; Nutt, SL; Yagita, H; Godfrey, DI; Hayakawa, Y. Sequential activation of NKT cells and NK cells provides effective innate immunotherapy of cancer. *J Exp Med*, 2005 201, 1973–1985.

[121] Spada, FM; Koezuka, Y; Porcelli, SA. CD1d-restricted recognition of synthetic glycolipid antigens by human natural killer T cells. *J Exp Med*, 1998 188, 1529–1534.

[122] Sriram, V; Cho, S; Li, P; O'Donnell, PW; Dunn, C; Hayakawa, K; Blum, JS; Brutkiewicz, RR. Inhibition of glycolipid shedding rescues recognition of a CD1[+] T cell lymphoma by natural killer T (NKT) cells. *Proc Natl Acad Sci USA*, 2002 99, 8197–8202.

[123] Stetson, DB; Mohrs, M; Reinhardt, RL; Baron, JL; Wang, ZE; Gapin, L; Kronenberg, M; Locksley, RM. Constitutive cytokine mRNAs mark natural killer (NK) and NK T cells poised for rapid effector function. *J Exp Med*, 2003 198, 1069–1076.

[124] Suzuki, Y; Wakita, D; Chamoto, K; Narita, Y; Tsuji, T; Takeshima, T; Gyobu, H; Kawarada, Y; Kondo, S; Akira, S; Katoh, H; Ikeda, H; Nishimura, T. Liposome-

encapsulated CpG oligodeoxynucleotides as a potent adjuvant for inducing type 1 innate immunity. *Cancer Res*, 2004 64, 8754–8760.

[125]   Tachibana, T; Onodera, H; Tsuruyama, T; Mori, A; Nagayama, S; Hiai, H; Imamura, M. Increased intratumor Va24-positive natural killer T cells: a prognostic factor for primary colorectal carcinomas. *Clin Cancer Res*, 2005 11, 7322–7327.

[126]   Tahir, SM; Cheng, O; Shaulov, A; Koezuka, Y; Bubley, GJ; Wilson, SB; Balk, SP; Exley, MA. Loss of IFN-gamma production by invariant NK T cells in advanced cancer. *J Immunol*, 2001 167, 4046–4050.

[127]   Takahashi, T; Nieda, M; Koezuka, Y; Nicol, A; Porcelli, SA; Ishikawa, Y; Tadokoro, K; Hirai, H; Juji, T. Analysis of human V alpha 24$^+$CD4$^+$ NKT cells activated by alpha-glycosylceramide-pulsed monocyte-derived dendritic cells. *J Immunol*, 2000 164, 4458–4464.

[128]   Taniguchi, M; Harada, M; Kojo, S; Nakayama, T; Wakao, H. The regulatory role of Va14 NKT cells in innate and acquired immune response. *Annu Rev Immunol*, 2003 21, 483–513.

[129]   Taniguchi, M; Seino, K; Nakayama, T. The NKT cell system: bridging innate and acquired immunity. *Nat Immunol*, 2003 4, 1164–1165.

[130]   Tatsumi, T; Takehara, T; Yamaguchi, S; Sasakawa, A; Sakamori, R; Ohkawa, K; Kohga, K; Uemura, A; Hayashi, N. Intrahepatic delivery of alpha-galactosylceramide-pulsed dendritic cells suppresses liver tumor. *Hepatology*, 2007 45, 22–30.

[131]   Terabe, M; Berzofsky; JA. Immunoregulatory T cells in tumor immunity. *Curr Opin Immunol*, 2004 16, 157–162.

[132]   Terabe, M; Khanna, C; Bose, S; Melchionda, F; Mendoza, A; Mackall, CL; Helman, LJ; Berzofsky, JA. CD1d-restricted natural killer T cells can down-regulate tumor immunosurveillance independent of interleukin-4 receptor-signal transducer and activator of transcription 6 or transforming growth factor-beta. *Cancer Res*, 2006 66, 3869–3875.

[133]   Terabe, M; Matsui, S; Noben-Trauth, N; Chen, H; Watson, C; Donaldson, DD; Carbone, DP; Paul, WE; Berzofsky, JA. NKT cell-mediated repression of tumor immunosurveillance by IL-13 and the IL-4R-STAT6 pathway. *Nature Immunol*, 2000 1, 515–520.

[134]   Terabe, M; Matsui, S; Park, J-M; Mamura, M; Noben-Trauth, N; Donaldson, DD; Chen, W; Wahl, SM; Ledbetter, S; Pratt, B; Letterio, JJ; Paul, WE; Berzofsky, JA. TGF-beta production and myeloid cell are an effector mechanism through which CD1d-restricted T cells block CTL-mediated tumor immunosurveillance: abrogation prevents tumor recurrence. *J Exp Med*, 2003 198, 1741–1752.

[135]   Terabe, M; Park, JM; Berzofsky, JA. Role of IL-13 in negative regulation of anti-tumor immunity. *Cancer Immunol Immunother* 2004 53, 79–85.

[136]   Terabe, M; Swann, J; Ambrosino, E; Sinha, P; Takaku, S; Hayakawa, Y; Godfrey, DI; Ostrand-Rosenberg, S; Smyth, MJ; Berzofsky, JA. A nonclassical non-Valpha14Jalpha18 CD1d-restricted (type II) NKT cell is sufficient for down-regulation of tumor immunosurveillance. *J Exp Med*, 2005 202, 1627–1633.

[137]   Tomura, M; Yu, WG; Ahn, HJ; Yamashita, M; Yang, YF; Ono, S; Hamaoka, T; Kawano, T; Taniguchi, M; Koezuka, Y; Fujiwara, H.. A novel function of Valpha14$^+$CD4$^+$ NKT cells: stimulation of IL-12 production by antigenpresenting cells in the innate immune system. *J Immunol*, 1999 163, 93–101.

[138] Toura, I; Kawano, T; Akutsu, Y; Nakayama, T; Ochiai, T; Taniguchi, M. Inhibition of experimental tumor metastasis by dendritic cells pulsed with alpha-galactosylceramide. *J Immunol*, 1999 163, 2387–2391.

[139] Tuyaerts, S; Aerts, JL; Corthals, J; Neyns, B; Heirman, C; Breckpot, K; Thielemans, K; Bonehill, A. Current approaches in dendritic cell generation and future implications for cancer immunotherapy. *Cancer Immunol Immunother*, 2007 56, 1513–1537.

[140] van den Broeke, LT; Daschbach, E; Thomas, EK; Andringa, G; Berzofsky, JA. Dendritic cell-induced activation of adaptive and innate antitumor immunity. *J Immunol*, 2003 171, 5842–5852.

[141] van der Vliet; Bakk, SB; Exley MA. Natural killer T cell-based cancer immunotherapy. *Clin Cancer Res*, 2006 12, 5921–5923.

[142] van der Vliet, HJ; Molling, JW; Blomberg, BM; Nishi, N; Kolgen, W; van den Eertwegh, AJ; Pinedo, HM; Giaccone, G; Scheper, RJ. The immunoregulatory role of CD1d-restricted natural killer T cells in disease. *Clin Immunol*, 2004 112, 8–23.

[143] van der Vliet, HJ; Molling, JW; Nishi, N; Masterson, AJ; Kölgen, W; Porcelli, SA; van den Eertwegh, AJ; von Blomberg, BM; Pinedo, HM; Giaccone, G; Scheper, RJ. Polarization of Valpha24R Vbeta11R natural killer T cells of healthy volunteers and cancer patients using alpha-galactosylceramide-loaded and environmentally instructed dendritic cells. *Cancer Res*, 2003 63, 4101–4106.

[144] van der Vliet, HJ; Nishi, N; Koezuka, Y; von Blomberg, BM; van den Eertwegh, AJ; Porcelli, SA; Pinedo, HM; Scheper, RJ; Giaccone, G. Potent expansion of human natural killer T cells using alpha-galactosylceramide (KRN7000)-loaded monocyte-derived dendritic cells, cultured in the presence of IL-7 and IL-15. *J Immunol Meth*, 2001 247, 61–72.

[145] Van Kaer, L. NKT cells: T lymphocytes with innate effector functions. *Curr Opin Immunol*, 2007 19, 354–364.

[146] Van Kaer, L. Regulation of immune responses by CD1d restricted natural killer T cells. *Immunol Res*, 2004 30, 139–153.

[147] Vela-Ojeda, J; Garcia-Ruiz Esparza, MA; Reyes-Maldonado, E; Jimenez-Zamudio, L; Garcia-Latorre, E; Moreno-Lafont, M; Estrada-Garcia, I; Mayani, H; Montiel-Cervantes, L; Tripp-Villanueva, F; Ayala-Sanchez, M; Garcia-Leon, LD; Borbolla-Escoboza, JR. Peripheral blood mobilization of different lymphocyte and dendritic cell subsets with the use of intermediate doses of G-CSF in patients with non-Hodgkin's lymphoma and multiple myeloma. *Ann Hematol*, 2006 85, 308–314.

[148] Vuckovic, S; Clark, GJ; Hart, DN. Growth factors, cytokines and dendritic cell development. *Curr Pharm Des*, 2002 8, 405–418.

[149] Weber, F; Byrne, SN; Le, S; Brown, DA; Breit, SN; Scolyer, RA; Halliday, GM. Transforming growth factor-beta1 immobilises dendritic cells within skin tumours and facilitates tumour escape from the immune system. *Cancer Immunol Immunother*, 2005 54, 898–906.

[150] Wilson, MT; Singh, AK; Van Kaer, L. Immunotherapy with ligands of natural killer T cells. *TRENDS Mol Med*, 2002 8, 225–231.

[151] Yanagisawa, K; Seino, K; Ishikawa, Y; Nozue, M; Todoroki, T; Fukao, K. Impaired proliferative response of Valpha24 NKT cells from cancer patients against alpha-galactosylceramide. *J Immunol*, 2002 168, 6494–6499.

[152] Yu, KO; Porcelli, SA. The diverse functions of CD1d-restricted NKT cells and their potential for immunotherapy. *Immunol Lett*, 2005 100, 42–55.

[153] Zhong, H; Shurin, MR; Han, B. Optimizing dendritic cell-based immunotherapy for cancer. *Expert Rev Vaccines*, 2007 6, 333–345.

[154] Zhou, D; Mattner, J; Cantu, C; Schrantz, N; Yin, N; Gao, Y; Sagiv, Y; Hudspeth, K; Wu, YP; Yamashita, T; Teneberg, S; Wang, D; Proia, RL; Levery, SB; Savage, PB; Teyton, L; Bendelac, A. Lysosomal glycosphingolipid recognition by NKT cells. *Science*, 2004 306, 1786–1789.

In: Natural Killer T-Cells: Roles, Interactions and Interventions   ISBN: 978-1-60456-287-3
Editor: Nathan V. Fournier, pp. 129-141   © 2008 Nova Science Publishers, Inc.

Chapter 5

Natural Killer T Cells in Peripheral Tolerance

Michael Nowak[*]

Beth Israel Deaconess Medical Center Harvard Medical School,
330 Brookline Ave., Boston, MA 02212

Abstract

Natural killer T (NKT) cells are innate immune cells that recognize lipid moieties presented by MHC -like CD1 molecules. The ability to secrete large amounts of cytokines rapidly upon activation has attracted considerable attention to their functions. Within the last decade it has become clear that NKT cells can steer immune responses towards either inflammation or tolerance. This ability is mostly attributed to the variety of different cytokines distinct NKT cell subsets can secrete and employ to generate a wide variety of effects on immunity, including control of infectious diseases, allergic responses, and cancer,, as well as tolerance in autoimmunity and transplantation settings. However, our understanding of the activation mechanisms of NKT cells suffers from the limited knowledge of the origin and identity of endogenous glycolipids that are responsible for the development and activation of NKT cells. The use of alpha-galactosylceramide (α-GalCer) an exogenous and semi-artificial glycolipid, elicits the secretion of various cytokines including those with antagonistic biological activities, leading to considerable confusion about which factors account for the effectof NKT cells in their different roles. This chapter summarizes current knowledge of the impact of NKT cells in different peripheral tolerance models and discusses possible mechanisms which control the biological outcomes, immunity or tolerance.

---

[*] E-mail address: mnowak1@bidmc.harvard.edu, Telephone: 001-617-667-4198.

Abbreviations

Anterior chamber associated immune deviation	(ACAID);
alpha galactosylceramide,	(α-GalCer);
antigen presenting cells	(APC);
dendritic cells	(DC);
experimental autoimmune encephalomyelitis	(EAE);
experimental myasthenia gravis	(EAMG);
graft versus host disease	(GVHD);
nonobese diabetic	(NOD);
insulin-dependent diabetes mellitus	(IDDM);
T cell receptor	(TCR);
T regulatory cell	(Treg).

Keywords: iNKT cell, peripheral tolerance, and autoimmunity

I. Introduction

NKT cells represent a subpopulation of T lymphocytes. The majority of NKT cells, termed invariant NKT (iNKT) cells, coexpress the glycoprotein NK1.1 together with a relatively invariant T cell receptor comprised of the alpha chain elements $V\alpha4J\alpha18$ in mice and the homologous elements $V\alpha24J\alpha Q$ in humans. The beta chain of their T cell receptor shows a biased association with $V\beta8.2$ element in mice [1] and $V\beta11$ in humans [2]. Whereas in humans iNKT cells can be divided into three subpopulations based on the expression of CD4 and CD8 molecules (double-negative, $CD4^+$, $CD8^+$) the latter subpopulation might not exist in mice. Despite general awareness of a heterogeneity of NKT cells [3], most of the work performed during the last decade since the discovery of NKT cells has focussed on iNKT cells; presumably this is due to lack of genetically engineered mouse strains and reagents for non-invariant NKT cells.

The majority of NKT cells, including iNKT cells, recognize glycolipids presented on the MHC class I-like monomorphic protein CD1 [4]. Upon activation, e.g. TCR cross-linking or recognition of glycolipids, iNKT cells rapidly secrete large amounts of a variety of cytokines, including IL-4, IL-10, IFN-γ, IL-13 [5, 6]. Although iNKT cells also can acquire cytotoxic effector functions through which they can directly elicit cell death in tumor cells [7], most attention has been attributed to their unique capacity to rapidly release different cytokines.

In a screening program for anti-tumor drugs alpha-galactosylceramide (α-GalCer), isolated from a marine sponge, was the first specific high-affinity CD1 ligand discovered. α-GalCer presentation to NKT cells in vivo and vitro induces the rapid and strong release of IL-4 and IFN-γ by iNKT cells [8]. Whereas a number of exogenous glycolipid ligands, derived from different bacterial strains such as *Mycobacterium*, *Ehrlichia* and *Sphingomonas*, have been identified [9]; our knowledge about endogenous ligands still is comparatively poor.

Within the last decade it has become clear that iNKT cells can steer immune responses towards Th1- or Th2-type immune responses [10]. Although much attention has been given to interactions between CD1-expressing antigen-presenting cells (APC), in particular dendritic cells (DC) ,[11-13], the regulation of cytokine secretion by iNKT cells and thus their effect on biological outcome of immune responses still is far from clear.

II. Cellular Interactions Regulating NKT Cell Development and Activation

The development of iNKT cells requires the expression of CD1. Although iNKT cell development has been suggested to occur in the liver[14] strong evidence exists that CD1-restricted NKT (iNKT) cells develop in the thymus [15, 16]. The expression of rodent CD1d molecules was first described in thymus, liver and spleen (reviewed in [17]). Indeed, the development of the majority of NKT cells depends on CD1 expressed in the thymus .Allthymocytes express CD1d molecules. Whereas the highest CD1d expression can be observed on cells of the $CD4^+CD8^+$ stage, CD4 and CD8 single-positive thymocytes also express this molecule. Upon thymic export peripheral T cells maintain low levels of CD1 expression; however, it is not clear whether they can act as antigen-presenting cells to NKT cells.

Albeit NK1.1⁻, a subset of murine $CD1^+$ cells carries an invariant T cell receptor, suggesting that they belong to NKT cells. This subset has been described to autopresent α-GalCer, implying that NKT cells can also act as antigen-presenting cells. It may be questioned, however, whether autopresentation of the superagonist α-GalCer reflects the presentation of low abundant endogenous or relevant exogenous glycolipids.

An important discovery was made by Zhou *et al.* [18] who identified the ceramide isoglobotrihexosylceramide (iGb_3) as a major endogenous ligand for NKT cells. Mice deficient in the β-subunit of hexosaminidase ($Hexb^{-/-}$ mice, a model for excessive lipid storage or Fabry's disease), which is involved in, but not critical for generation of iGb_3, showed largely diminished NKT cell numbers. From these results the authors concluded presentation of iGb_3 to be required for the NKT cell development In contrast, mice deficient in the rate-limiting enzyme for iGb_3 production, iGb_3 synthase, showed no numerical or functional alterations of NKT cells. Additionally, the authors of an accompanying paper could not provide any biochemical evidence for iGb_3 in the thymus [19]. Most likely, an altered loading of CD1 molecules with accumulating and competing lipids in $Hexb^{-/-}$ mice accounted for the observed NKT cell alterations.

Dendritic cells are the most potent activators of NKT cells. Extensive body of evidence exists that costimulatory molecules have a decisive role for the differential activation of NKT cells (i.e. their cytokine secretion). Hayakawa *et al.* have dissected the role of costimulatory molecules in this process demonstrating that blockade of either the CD28:CD80/86 or CD154:CD40 axis abrogates the production of IFNγ by NKT cells, whereas inhibition of IL-4 secretion in response to α-GalCer recognition required the blockade of both signalling pathways [20]. These data suggest that secretion of pro-inflammatory cytokines (e.g. IFNγ) by NKT cells is more tightly regulated than the secretion of anti-inflammatory/ tolerance-promoting cytokines like IL-4. A decisive role for the control of IFNγ secretion was attributed to IL-12 secreted from antigen-presenting cells (APCs) in response to CD40 ligation. α-GalCer-pulsed hepatocytes which express CD1 but lack costimulatory molecules were able to stimulate IFNγ secretion in NKT cells in vitro only when exogenous IL-12 was added to the cultures, whereas liver DCs expressing lower amounts of CD1 compared to hepatocytes triggered the secretion of both IL-4 and IFNγ [21]. Given the fact that both cell types still secreted IL-4 upon ligation, solely the lack of costimulatory molecules on hepatocytes, as suggested by the study discussed above [20] cannot account for the predominant IL-4 secretion observed.Conceivably, a possible explanation for the differential cytokine response lies in a differential avidity of their ligand:TCR interactions.

Recently, Chen *et al.* reported that, independently of IL-12, NKT responses to dendritic cells can be amplified by recognition of endogenous glycolipid [22]. DCs were found capable amplifying the NKT cell response to exogenous antigens, such as α-GalCer, by presenting the endogenous lipid ligand iGb$_3$. NKT cell activation occurred independently of whether the exogenous (i.e. α-GalCer) and endogenous glycolipid (i.e. iGb$_3$) were presented by the same APC. It is tempting to speculate that different DC subsets might be able to also amplify NKT cell responses to low abundant self glycolipids. This notion is supported by findings of Wiethe *et al.* who showed that murine bone-marrow derived DCs stimulated by lipopolysaccharide (LPS) plus CD40 preferentially stimulated IFNγ secretion in NKT cells whereas DC maturation induced by exogenous TNF led to DCs that stimulated IL-4 responses [23]. These findings suggest that the context in which APCs are activated/matured decide about the outcome of NKT responses rather than the type of antigen (i.e. glycolipid) presented.

In view of the relative broad spectrum of non-professional APCs that express CD1 in the periphery it is conceivable that DCs may also amplify and regulate responses of NKT cells activated in the periphery (e.g. the CNS).

III. Impact of NKT Cells on Autoimmunity and Tolerance

The ability of the immune system to discriminate between self and foreign antigens is the basis for immune tolerance. The latter is acquired by a variety of different mechanisms. Potentially autoreactive T cells are clonally eliminated during their thymic development, a process called central tolerance. Since a large number of autoreactive T cells escapes this process, mechanisms in the periphery serve to delete or inhibit these cells by induction of either cell death, anergy, or immune deviation through cytokines [24-26]; failure in these mechanisms can result in autoimmunity. The contribution of different subsets of regulatory T (Treg) cells and dendritic cells to peripheral tolerance has extensively been studied [27, 28].

A large body of evidence indicates that iNKT cells as well are involved in the regulation of autoimmunity and maintenance of self-tolerance. Numerical deficiencies of iNKT cells have been observed in patients and mice with different autoimmune diseases, including but not limited to type-I diabetes (T1D) [29], multiple sclerosis (MS)[30], and experimental autoimmune encephalomyelitis (EAE) [31]. Conversely, experimentally induced overproduction of NKT cells in susceptible mouse strains ameliorates autoimmune diseases or delays their onset [32].

Type-1 Diabetes

The impact of NKT cells on the prevention and experimental therapy probably is best understood in type-1 diabetes (T1D).

Insulin-dependent diabetes mellitus is an autoimmune disease that spontaneously develops in non-obese diabetes (NOD) mice and is characterized by the CD8-T cell mediated β-cell islet destruction [33]. Both T1D patients and NOD mice, a strain spontaneously developing T1D, exhibit decreased numbers of NKT cells. A functional impact of NKT cells on the prevention of diabetes in NOD mice was derived from the early observation that reconstitution of young NOD mice with NKT cells prevented the diabetes development [34]. Likewise the crossing of NOD mice to mice overexpressing iNKT cells ameliorated the disease onset [35]. Reversely, crossing of the CD1-null gene, which prevents NKT cell development, onto NOD genetic background accelerated the onset of diabetes [36].

The exact mechanisms by which NKT cells inhibit autoimmune attacks are only incompletely understood. Studies in NOD suggest that NKT cells act in a contact-dependent fashion. Chen *et al.* demonstrated that activation of NKT cells by α-GalCer administration leads to DC maturation in the pancreatic lymph nodes which subsequently resulted in influx of self-reactive T cells and subsequently their anergy / apoptosis induction [37]. The activity of NKT cells in T1D depends on the functional activity of $CD4^+CD25^+$ Tregs since diabetes protection by α-GalCer administration could only be achieved when Tregs were present. Conversely, the Treg activity was independent of presence or activation status of NKT cells [38]. These findings suggest that NKT cells are needed in the induction phase of $CD4^+CD25^+$ Tregs , but not in their effector phase. This notion is further supported by data obtained in models of anterior-chamber of the eye induced immune deviation (ACAID,) and oral tolerance to contact allergen nickel, as will discussed further below. Whether the administration of the endogenous super-agonist α-GalCer reflects mechanisms in the steady-state remains to be clarified. Which endogenous glycolipids are recognized by NKT cells in the steady-state to prevent the onset of diabetes and whether presentation of such glycolipids by dendritic cells is operational is not known.

Myasthenia Gravis

Evidence for interactions between NKT and Treg cells has been obtained from studies using experimental autoimmune myasthenia gravis (EAMG), the animal model for human myasthenia gravis (MG). Both EAMG and MG are T cell-driven autoimmune reactions directed against the actylcholine receptor (AChR) resulting in neuromuscular disorders [39].

NKT cells did not contribute to the prevention of EAMG in the steady-state, that is in the absence of exogenous glycolipid administration since wild-type C57BL/6, $J\alpha18^{-/-}$ mice and $CD1^{-/-}$ mice, respectively, immunized with AChR showed comparable disease progression, activation of NKT cells by preadministration of α-GalCer protected WT mice from the onset of EAMG. This protection was accompanied with Treg cell proliferation that was mediated by IL-2 secreted by NKT cells [40]. Whether in this experimental model the subset of $CD4^+$ NKT cells are instrumental for the IL-2 secretion needs to be clarified. EAMG protection in this experimental setup proved independent of IL-4 secreted by NKT cells[40] underscoring the notion that iNKT cells can promote tolerance by mechanisms other than IL-4 / IFNγ mediated immune deviation.

The findings that NKT cells in different tolerance models, at least partially by IL-2 secretion, can promote Treg expansion should facilitate further search of CD1 ligands that elicit particularly the secretion of IL-2.

Experimental Autoimmune Encephalomyelitis

Experimental autoimmune encephalomyelitis (EAE), the rodent model of multiple sclerosis (MS). Murine EAE can be elicited by injecting myelin oligodendrocyte glycoprotein (MOG) antigen, susceptible mouse strains subsequently show infiltrations of granulocytes, macrophages and T cells into the CNS causing edema and (reversible) paralysis. MS and EAE are characterized by alternating cycles of disease relapse and remisson which are accompanied by waves of predominant secretion of Th1 and Th2 cytokines, respectively, produced by lymphocytes including NK and NKT cells [41]. Thus, CD4$^+$ iNKT cells from patients with MS in relapse exhibit a decreased IL-4 production [42], an observation consistent to findings in other autoimmune diseases [43, 44]. The administration of α-GalCer, however, has led to paradoxical results. Activation of NKT cells using α-GalCer prior to immunization with the encephalitogenic peptide resulted in prompt release of IFNγ. This was followed by inhibition of IL-4 production and thus, unexpectedly, resulted in a Th2 bias in pathogenic effector T cells and relapse[45]. Conversely, another study employed NKT cell activation subsequent to immunization leading to excacerbation of EAE[46]. Here, the *excacerbation* of EAE was attributed to α-GalCer-driven NKT cell-mediated IFNγ. Comparable to other autoimmune models, administration of α-GalCer resulted in EAE excacerbation mediated by the production of IL-4 and IL-10 by iNKT cells [45].

Although invariant NKT cells in EAE have attracted much attention additional subsets of NKT cells might participate in disease control. Non-invariant CD1-restricted NKT cells distinct from iNKT cells accumulate in the CNS of EAE mice and recognize the myelin-derived glycolipid sulfatide as an endogenous ligand [46]. Interestingly, the administration of this sulfatide to WT but not CD1-null mice resulted in inhibition of both IL-4 and IFNγ responses by myelin-reactive conventional T cells rather than a shift towards Th2 cytokine secretion mediated by non-invariant NKT cells. Whether non-invariant NKT cells described in this study control the activation of iNKT cells is not clear.

Taken together, the multitude of results in EAE obtained using different protocols of α-GalCer administration strongly suggest more complex but yet elusive mechanisms underlying the NKT-mediated effects in tolerance than just a proposed simplistic Th1/Th2 deviation.

The involvement of a further endogenous glycolipidin EAE protection has recently been proposed. Wiethe *et al.* [23] suggested that DCs simultaneously present MOG-derived peptides to conventional T cells the, yet unidentified, endogenous glycolipidto iNKT cells which leads to NKT cell-derived production of IL-4 and IL-4 / IL-10 produced by conventional T cells, leading to EAE protection. Treatment of the DCs with isolectin B$_4$, a lectin that was reported to selectively block the recognition of iGb$_3$ blocked this effect, suggesting that this endogenous glycolipid might be involved in this process.

Whether NKT cells regulate DCs to maintain a semi-mature state and thus inhibit priming of autoreactive T cells remains to be clarified. An alternative, but not exclusive mechanism to this is that DCs might license NKT cells to produce Th2 cytokines which in turn might directly inhibit the priming of Th1 effector T cells. None of these findings exclude the possibility that endogenous glycolipids presented by non-professional APCs of the CNS, including Schwann cells or microglia, contribute to maintain NKT cells in a tolerance-promoting phenotype in the state of health. Schwann cells express CD1 molecules and were shown to be able to activate iNKT cells when pulsed with α-GalCer [47]. Noteworthy, Schwann cells preferentially activate NKT cells to secrete Th2 cytokines, could not be reverted by overexpressing CD1 molecules which are normally expressed at low levels on Schwann cells. The predominant Th2-activating capacity was ascribed to the lack of IL-12 production. However, the addition of exogenous IL-12 only marginally increased IFNγ of NKT cells and failed to inhibit the secretion of IL-5 and IL-13 suggesting the Th2-activating capacity is might be intrinsic to Schwann cells.

However, cells of the immune system patroling the CNS as well as resident microglia under physiological conditions exhibit a markedly reduced CD1 expression that is upregulated in EAE lesions [48]. This might suggest that under healthy conditions the recognition of glycolipids (of yet unknown identity) is suppressed by the elevated recognition threshold rather than deviated to presentation of tolerogenic signal.

Anterior-Chamber of the Eye Induced Immune Deviation (ACAID)

The regulatory function of NKT cells for induction of immune tolerance has first been described and most thoroughly investigated up to now in the ACAID model [49]. Here, ocular injected antigen, such as ovalbumin (OVA), induces systemic tolerance. OVA thus administered is transported to the splenic marginal zone (MZ) by eye-derived CD1d$^+$ F4/80$^+$ APCs. CD4$^+$ NKT cells are attracted to the MZ where they aggregate with both F4/80$^+$ APCs and MZ B cells [50, 51]. The interaction between CD1 on F4/80$^+$APC and iNKT cells is necessary for their activation and production of the chemokine RANTES which recruits further F4/80$^+$ cells and T cells into the MZ. [52]. Since iNKT cells produce cytokines IL-10 and TGFβ [50] a role of NKT cells in the development of regulatory T cells has been attributed. Interestingly, and different to the above discussed models of EAE and diabetes is, the endpoint of a clustering between the F480$^+$APC, MZ B, and NKT cells is the development of efferent CD8$^+$ Tregs. A further important difference to other tolerance models is that ACAID is induced independently of the IL-4 production (e.g, by iNKT cells).

To date no data are available on the identity of the glycolipids presented by the F4/80$^+$ APCs and MZ B cells. One may speculate whether spleen-resident MZ B cells may amplify the signals delivered from the eye-derived F4/80$^+$ APC to NKT cells. Such a feasible function would be comparable to the signal amplification to low abundant exogenous glycolipids by presentation of the self-glycolipid iGb$_3$ on DCs ([22], cf. chapter II). Whether activation of NKT cells in this system involves presentation of iGb$_3$ as suggested in the model of experimental autoimmune encephalomyelitis (EAE, discussed below) has to wait

Oral Tolerance to Nickel

Further support for the concept of NKT cell-mediated help for Tregs has been obtained from studies in mice using the contact allergen nickel. Oral administration of Ni^{2+} ions generates Ni-specific neoantigens [53] and induces long-lasting systemic tolerance that is accompanied by expansion of a subset of $\alpha_E\beta_{7+}CD25^+$ T cells [54]. In particular, a subset of CD4$^+$ NKT cells elicited FasLigand-mediated apoptosis in Ni-carrying CD1d$^+$ B cells. Apoptotic nickel-neoantigen containing cellular corpses are engulfed by dendritic cells and subsequently presented, leading to the development of Treg cells. B cells in this system directly activate NKT cells to excert their cytotoxic function [55]. How B cells activate NKT cells and trigger their own apoptosis induction is not fully understood. Conceivably, effete B cells damaged by the genotoxic activity of nickel signal activate NKT cells by presentation of a yet unknown endogenous glycolipid. iNKT cells in that model produced the cytokines IL-4 and IL-10 which were required for the generation of Tregs.Unlike in other models discussed above, IL-4 and IL-10 production did not result in immune deviation but instead proved to be involved upstream of the NKT effector functions, most likely the upregulation of cytotoxic molecule FasLigand.

The data obtained in the system of oral tolerance to nickel support the idea that NKT cells can promote tolerance regulating the development of Treg cells but act independently of cytokine-mediated immune deviation. Unfortunately, this fact has often been disregarded in studies employing activation of NKT cells using the super-agonist α-GalCer and, according to the present author, may have led to some misconceptions about the orchestration of tolerance. Whether CD4$^+$NKT cells directly (i.e. IL-2 secretion) promote Treg expansion, as described in the system of EAMG [40] needs to be tested.

Conceivably, but not mutually exclusive to that possibility, is that NKT cells might indirectly support the Treg development by eliciting apoptosis in antigen-carrying B cells and thus providing significant amounts of self-antigen that can be presented on DCs to T cells.

Conclusion

NKT cells possess the remarkable capacity to steer immune responses by providing large amounts of different cytokines, chemotactic stimuli as well as cytotoxic functions.

Admittedly, the use of the high-affinity ligand α-GalCer has greatly expedited the functions of NKT cells and their impact in different immune responses. On the other hand, however, results obtained after administration of this exogenous glycolipid unilaterally focussed on cytokine secretion and thus generated skewed concepts in the field. A central question remaining is how NKT cells support tolerance under more physiological conditions, a question NKT cell research should try to answer in the future.

Questions Regarding the Control of NKT Cell Functions

- What is the impact of the chemical nature of glyclipids presented on the outcome of NKT cell activation?
- What is the impact of different NKT cell subsets (CD4$^+$ / double-negative) and their organ distribution for the cytokine release?
- Can we identify glycolipids eliciting the exclusive secretion of IL-2 and IL-4 or IL-10 by NKT cells?
- Are we able to identify endogenous lipids eliciting tolerogenic NKT functions?

Table 1. Mode of action of NKT cells in tolerance models and autoimmune diseases

Tolerance model / autoimmune disease	Type of APC known / suggested to be involved in NKT activation	Subset of NKT cells involved	Interaction of NKT cells with other cell types /Effector function of NKT cells
Type-I diabetes	Dendritic cells	CD4−CD8− NKT cells[56]	Cytokine production (IL-4)[35] α-GalCer activated iNKT cells recruit DC and induce maturation in pancreatic LN[37]
Multiple sclerosis / EAE	Schwann cells[47] Microglia[48] Dendritic cells[23]	Non-invariant CD1-restricted NKT cells[46]	EAE: α-GalCer activated NKT cells secrete IL-4 and IL-10[45] Non-invariant NKT cells inhibit cytokine secretion of T cells in response to sulfatide administration[46]
Myasthenia gravis / EAMG	Unknown	Unknown	NKT-derived IL-2 induces Treg proliferation[40]
ACAID	Cell clusters in splenic MZ contain eye-derived, antigen-carrying F4/80+ APC and MZ B cells[51]	CD4+ iNKT [57]	NKT-derived chemokines recruit T cells and APC into splenic marginal zone [52], IL-10 production [50]
Oral tolerance to nickel	Splenic B cells[55]	CD4+ iNKT [55, 58]	IL-4 and IL-10 production[58] and FasLigand-mediated killing by splenic iNKT cells of antigen-carrying B cells[55]

- What is the impact of the activation/maturation status of APCs for the different NKT effector functions?
- What is the biological outcome of glycolipid recognition on non-professional antigen-presenting cells ? Does recognition on such cells promote tolerance?
- By which mechanisms promote NKT cells tolerance in the steady-state, i.e. without exogenous activation?

References

[1]   Hayakawa K, Lin BT, Hardy RR. Murine thymic CD4+ T cell subsets: a subset (Thy0) that secretes diverse cytokines and overexpresses the V beta 8 T cell receptor gene family. *J Exp Med* 1992;176(1):269-74.

[2]   Bendelac A, Rivera MN, Park SH, Roark JH. Mouse CD1-specific NK T cells: development, specificity, and function. *AnnuRevImmunol* 1997;15:535-62.

[3]   Emoto M, Kaufmann SH. Liver NKT cells: an account of heterogeneity. *Trends Immunol* 2003;24(7):364-9.

[4]   Godfrey DI, McCluskey J, Rossjohn J. CD1d antigen presentation: treats for NKT cells. *NatImmunol* 2005;6(8):754-6.

[5]   Terabe M, Matsui S, Noben-Trauth N, *et al.* NKT cell-mediated repression of tumor immunosurveillance by IL-13 and the IL-4R-STAT6 pathway. NatImmunol 2000;1(6):515-20.

[6]   Sieling PA, Chatterjee D, Porcelli SA, *et al.* CD1-restricted T cell recognition of microbial lipoglycan antigens. *Science* 1995;269(5221):227-30.

[7]   Nicol A, Nieda M, Koezuka Y, Porcelli S, Suzuki K, Tadokoro K, Durrant S, Juji T. Human invariant valpha24+ natural killer T cells activated by alpha-galactosylceramide (KRN7000) have cytotoxic anti-tumour activity through mechanisms distinct from T cells and natural killer cells. *Immunology* 2000;99(2):229-34.

[8]   Morita M, Motoki K, Akimoto K, *et al.* Structure-activity relationship of alpha-galactosylceramides against B16-bearing mice. JMedChem 1995;38(12):2176-87.

[9]   Bendelac A, Savage PB, Teyton L. The biology of NKT cells. *AnnuRevImmunol* 2007;25:297-336.

[10]   Nowak M, Stein-Streilein J. Invariant NKT cells and tolerance. *IntRevImmunol* 2007;26(1-2):95-119.

[11]   Naumov YN, Bahjat KS, Gausling R, *et al.* Activation of CD1d-restricted T cells protects NOD mice from developing diabetes by regulating dendritic cell subsets. *ProcNatlAcadSciUSA* 2001;98(24):13838-43.

[12]   Ikarashi Y, Mikami R, Bendelac A, *et al.* Dendritic cell maturation overrules H-2D-mediated natural killer T (NKT) cell inhibition: critical role for B7 in CD1d-dependent NKT cell interferon gamma production. *JExpMed* 2001;194(8):1179-86.

[13]   Kadowaki N, Antonenko S, Ho S, Rissoan MC, Soumelis V, Porcelli SA, Lanier LL, Liu YJ. Distinct cytokine profiles of neonatal natural killer T cells after expansion with subsets of dendritic cells. *JExpMed* 2001;193(10):1221-6.

[14]   Sato K, Ohtsuka K, Hasegawa K, Yamagiwa S, Watanabe H, Asakura H, Abo T. Evidence for extrathymic generation of intermediate T cell receptor cells in the liver revealed in thymectomized, irradiated mice subjected to bone marrow transplantation. *JExpMed* 1995;182(3):759-67.

[15]   Coles MC, Raulet DH. NK1.1+ T cells in the liver arise in the thymus and are selected by interactions with class I molecules on CD4+CD8+ cells. *JImmunol* 2000;164(5) 2412-8.

[16]   Hammond K, Cain W, van DI, Godfrey D. Three day neonatal thymectomy selectively depletes NK1.1+ T cells. *IntImmunol* 1998;10(10):1491-9.

[17]   Dougan SK, Kaser A, Blumberg RS. CD1 expression on antigen-presenting cells. Curr *Top Microbiol Immunol* 2007;314:113-41.

[18]   Zhou D, Mattner J, Cantu C, III, *et al.* Lysosomal glycosphingolipid recognition by NKT cells. *Science* 2004;306(5702):1786-9.

[19]   Porubsky S, Speak AO, Luckow B, Cerundolo V, Platt FM, Grone HJ. Normal development and function of invariant natural killer T cells in mice with isoglobotrihexosylceramide (iGb3) deficiency. *Proc Natl Acad Sci U S A* 2007;104(14):5977-82.

[20]   Hayakawa Y, Takeda K, Yagita H, Van Kaer L, Saiki I, Okumura K. Differential regulation of Th1 and Th2 functions of NKT cells by CD28 and CD40 costimulatory pathways. *JImmunol* 2001;166(10):6012-8.

[21]   Trobonjaca Z, Leithauser F, Moller P, Schirmbeck R, Reimann J. Activating immunity in the liver. I. Liver dendritic cells (but not hepatocytes) are potent activators of IFN-gamma release by liver NKT cells. *JImmunol* 2001;167(3):1413-22.

[22]   Cheng L, Ueno A, Cho S, Im JS, Golby S, Hou S, Porcelli SA, Yang Y. Efficient activation of Valpha14 invariant NKT cells by foreign lipid antigen is associated with concurrent dendritic cell-specific self recognition. *JImmunol* 2007;178(5):2755-62.

[23]   Wiethe C, Schiemann M, Busch D, Haeberle L, Kopf M, Schuler G, Lutz MB. Interdependency of MHC class II/self-peptide and CD1d/self-glycolipid presentation by TNF-matured dendritic cells for protection from autoimmunity. *JImmunol* 2007;178(8):4908-16.

[24]   Kisielow P, Bluthmann H, Staerz UD, Steinmetz M, von Boehmer H. Tolerance in T-cell-receptor transgenic mice involves deletion of nonmature CD4 + CD8 + thymocytes *Nature* 1988;333(6175):742-6.

[25]   Lanoue A, Bona C, von Boehmer H, Sarukhan A. Conditions that induce tolerance in mature CD4 + T cells. *JExpMed* 1997;185(3):405-14.

[26]   Jordan MS, Riley MP, von Boehmer H, Caton AJ. Anergy and suppression regulate CD4(+) T cell responses to a self peptide. *EurJImmunol* 2000;30(1):136-44.

[27]   Sakaguchi S. Naturally Arising CD4+ Regulatory T Cells for Immunologic Self-Tolerance and Negative Control of Immune Responses. *AnnuRevImmunol* 2004 22:531-62.

[28]   Steinman RM, Nussenzweig MC. Avoiding horror autotoxicus: the importance of dendritic cells in peripheral T cell tolerance. *ProcNatlAcadSciUSA* 2002;99(1):351-8.

[29]   Poulton LD, Smyth MJ, Hawke CG, Silveira P, Shepherd D, Naidenko OV, Godfrey DI, Baxter AG. Cytometric and functional analyses of NK and NKT cell deficiencies in NOD mice. *IntImmunol* 2001;13(7):887-96.

[30] Sumida T, Sakamoto A, Murata H, *et al.* Selective reduction of T cells bearing invariant V alpha 24J alpha Q antigen receptor in patients with systemic sclerosis. *JExpMed* 1995;182(4):1163-8.

[31] Kojo S, Adachi Y, Keino H, Taniguchi M, Sumida T. Dysfunction of T cell receptor AV24AJ18+, BV11+ double-negative regulatory natural killer T cells in autoimmune diseases. *Arthritis Rheum* 2001;44(5):1127-38.

[32] Lehuen A, Lantz O, Beaudoin L, Laloux V, Carnaud C, Bendelac A, Bach JF, Monteiro RC. Overexpression of natural killer T cells protects Valpha14- Jalpha281 transgenic nonobese diabetic mice against diabetes. *JExpMed* 1998;188(10):1831-9.

[33] Makino S, Kunimoto K, Muraoka Y, Mizushima Y, Katagiri K, Tochino Y. Breeding of a non-obese, diabetic strain of mice. Jikken Dobutsu 1980;29(1):1-13.

[34] Hammond KJ, Poulton LD, Palmisano LJ, Silveira PA, Godfrey DI, Baxter AG. alpha/beta-T cell receptor (TCR) + CD4 - CD8 - (NKT) thymocytes prevent insulin-dependent diabetes mellitus in nonobese diabetic (NOD)/Lt mice by the influence of interleukin (IL)-4 and/or IL-10. *JExpMed* 1998;187(7):1047-56.

[35] Laloux V, Beaudoin L, Jeske D, Carnaud C, Lehuen A. NK T cell-induced protection against diabetes in V alpha 14-J alpha 281 transgenic nonobese diabetic mice is associated with a Th2 shift circumscribed regionally to the islets and functionally to islet autoantigen. *JImmunol* 2001;166(6):3749-56.

[36] Shi FD, Flodstrom M, Balasa B, Kim SH, Van Gunst K, Strominger JL, Wilson SB, Sarvetnick N. Germ line deletion of the CD1 locus exacerbates diabetes in the NOD mouse. *ProcNatlAcadSciUSA* 2001;98(12):6777-82.

[37] Chen YG, Choisy-Rossi CM, Holl TM, *et al.* Activated NKT cells inhibit autoimmune diabetes through tolerogenic recruitment of dendritic cells to pancreatic lymph nodes. *JImmunol* 2005;174(3):1196-204.

[38] Ly D, Mi QS, Hussain S, Delovitch TL. Protection from type 1 diabetes by invariant NK T cells requires the activity of CD4+CD25+ regulatory T cells. *J Immunol* 2006;177(6):3695-704.

[39] Vincent A. Unravelling the pathogenesis of myasthenia gravis. *NatRevImmunol* 2002;2(10):797-804.

[40] Liu R, La Cava A, Bai XF, *et al.* Cooperation of invariant NKT cells and CD4+CD25+ T regulatory cells in the prevention of autoimmune myasthenia. *J Immunol* 2005;175(12):7898-904.

[41] Steinman L. Multiple sclerosis: a two-stage disease. *NatImmunol* 2001;2(9):762-4.

[42] Gausling R, Trollmo C, Hafler DA. Decreases in interleukin-4 secretion by invariant CD4(-)CD8(-)V alpha 24J alpha Q T cells in peripheral blood of patientswith relapsing-remitting multiple sclerosis. *ClinImmunol* 2001;98(1):11-7.

[43] Gombert JM, Herbelin A, Tancrede-Bohin E, Dy M, Carnaud C, Bach JF. Early quantitative and functional deficiency of NK1+-like thymocytes in the NOD mouse. *EurJImmunol* 1996;26(12):2989-98.

[44] Mieza MA, Itoh T, Cui JQ, *et al.* Selective reduction of V alpha 14+ NK T cells associated with disease development in autoimmune-prone mice. *J Immunol* 1996;156(10):4035-40.

[45] Furlan R, Bergami A, Cantarella D, Brambilla A, Taniguchi M, Belladonna P, Casorati G, Martino G. Activation of invariant NKT cells by alphaGalCer

administration protects mice from MOG35-55-induced EAE: critical roles for administration route and IFN-gamma. *EurJImmunol* 2003;33(7):1830-8.

[46]   Jahng AW, Maricic I, Pedersen B, Burdin N, Naidenko O, Kronenberg M, Koezuka Y, Kumar V. Activation of natural killer T cells potentiates or prevents experimental autoimmune encephalomyelitis. *JExpMed* 2001;194(12):1789-99.

[47]   Im JS, Tapinos N, Chae GT, *et al.* Expression of CD1d molecules by human schwann cells and potential interactions with immunoregulatory invariant NK T cells. *JImmunol* 2006;177(8):5226-35.

[48]   Busshoff U, Hein A, Iglesias A, Dorries R, Regnier-Vigouroux A. CD1 expression is differentially regulated by microglia, macrophages and T cells in the central nervous system upon inflammation and demyelination. *JNeuroimmunol* 2001;113(2):220-30.

[49]   Stein-Streilein J, Streilein JW. Anterior chamber associated immune deviation (ACAID): regulation, biological relevance, and implications for therapy. *IntRevImmunol* 2002;21(2-3):123-52.

[50]   Sonoda KH, Faunce DE, Taniguchi M, Exley M, Balk S, Stein-Streilein J. NK T cell-derived IL-10 is essential for the differentiation of antigen-specific T regulatory cells in systemic tolerance. *JImmunol* 2001;166(1):42-50.

[51]   Sonoda KH, Stein-Streilein J. CD1d on antigen-transporting APC and splenic marginal zone B cells promotes NKT cell-dependent tolerance. *EurJImmunol* 2002;32(3):848-57.

[52]   Faunce DE, Stein-Streilein J. NKT cell-derived RANTES recruits APCs and CD8 + T cells to the spleen during the generation of regulatory T cells in tolerance. *JImmunol* 2002;169(1):31-8.

[53]   Lu L, Vollmer J, Moulon C, Weltzien HU, Marrack P, Kappler J. Components of the ligand for a Ni ++ reactive human T cell clone. *JExpMed* 2003;197(5):567-74.

[54]   Wu X, Roelofs-Haarhuis, K., Zhang, J., Nowak, M., Layland, L., Jermann, E., Gleichmann, E. Dose dependence of oral tolerance to nickel. *IntImmunol* (in press) 2007.

[55]   Nowak M, Kopp, F., Roelofs-Haarhuis, K. Wu, X., Gleichmann, E. Oral Nickel Tolerance: Fas Ligand-Expressing Invariant NK T Cells Promote Tolerance Induction by Eliciting Apoptotic Death of Antigen-Carrying, Effete B Cells. *JImmunol* 2006;176(8):4581-9.

[56]   Baxter AG, Kinder SJ, Hammond KJ, Scollay R, Godfrey DI. Association between alphabetaTCR+CD4-CD8- T-cell deficiency and IDDM in NOD/Lt mice. *Diabetes* 1997;46(4):572-82.

[57]   Nakamura T, Sonoda KH, Faunce DE, Gumperz J, Yamamura T, Miyake S, Stein-Streilein J. CD4+ NKT Cells, But Not Conventional CD4+ T Cells, Are Required to Generate Efferent CD8+ T Regulatory Cells Following Antigen Inoculation in an Immune-Privileged Site. *JImmunol* 2003;171(3):1266-71.

[58]   Roelofs-Haarhuis K, Wu X, Gleichmann E. Oral tolerance to nickel requires CD4(+) invariant NKT cells for the infectious spread of tolerance and the induction of specific regulatory T cells. *JImmunol* 2004;173(2):1043-50.

In: Natural Killer T-Cells: Roles, Interactions and Interventions   ISBN: 978-1-60456-287-3
Editor: Nathan V. Fournier, pp. 143-153   © 2008 Nova Science Publishers, Inc.

Chapter 6

Possible Clinical Use of NK Cells for the Control of Hematologic Malignancies

Giovanni F. Torelli, Roberta Maggio, Maria G. Mascolo,
Filippo Milano, Barbarella Lucarelli, Walter Barberi,
Veronica Valle, Emilia Iannella, Anna P. Iori,
Anna Guarini and Robin Foà

Division of Hematology, Department of Cellular Biotechnologies and Hematology,
University "La Sapienza", Rome, Italy

Abstract

The anti-leukemic potential of natural killer (NK) cells and their competence in regulating normal and possibly neoplastic hematopoietic precursors have over the years raised considerable interest. The role of NK cells in the immunosurveillance against tumor growth is well documented. Previous studies have shown that leukemic blasts may be susceptible to the lytic action of lymphokine-activated killer cells. More recently, NK clones of donor origin have been established in the post-transplant period from HLA-mismatched hematopoietic stem cell transplanted patients. NK clones were capable of killing recipients' leukemic cells, in the absence of graft-versus-host disease (GVHD). We have recently demonstrated the possibility of expanding cytotoxic NK cells with killing activity against autologous and allogeneic blasts from acute lymphoid leukemia (ALL) and acute myeloid leukemia (AML) patients in complete remission (CR). The cytolytic properties of this expanded cell population have been further confirmed *in vivo* in a NOD/SCID mouse tumor model that showed a consistent reduction of AML load after adoptive transfer of autologous expanded interleukin (IL)-2- and IL-15-activated NK cells into tumor-transplanted mice. These results are of particular interest if we consider the high rate of relapse that characterizes the clinical course of leukemia patients. These effectors expanded *ex vivo* may be used for vaccination programs aimed at controlling or eradicating minimal residual disease in leukemic patients in clinical and hematologic CR. Alternatively, this population of cytotoxic cells may be utilized in the setting of allografted patients, both in the haploidentical or HLA-matched scenario. Clinical protocols based on autologous or allogeneic NK cells expanded under Good Manufacturing Practice conditions appear feasible, particularly considering that the infusion of NK cells should induce very limited toxicity and no or a very low risk of GVHD.

Introduction

After many years of disillusion, the possibility of utilizing immune-mediated approaches to control neoplastic clones has become a reality in various hematologic malignancies. This is largely a consequence of the continuous advances in knowledge and of the progressive development of more refined technologies that have led to a better understanding of the biology of the malignant cells and of the host immune system, and to the design of innovative therapeutic programs. The general rationale behind immunotherapeutic strategies is the hope that cells potentially capable of exerting an anti-neoplastic effect could allow a more physiologic control of cancer cells, particularly in the setting of minimal residual disease (MRD).

The observation that after allogeneic stem cell transplantation (SCT) the graft-versus-leukemia (GVL) response to residual disease is based on immunologic mechanisms, represented the first real proof of the potential impact of therapeutic approaches based on immunologic recognition [1,2]; the success of therapeutic donor leukocyte infusion in the treatment of relapse definitively confirmed this hypothesis [3-6].

Based on these considerations, several direct or indirect strategies aimed at an immunologic control of the disease have shown activity in different hematologic conditions or are starting to reach the clinical practice. Different approaches are based on the utilization of the various subpopulations of lymphoid cells. One of them are natural killer (NK) cells, whose competence in controlling neoplastic precursours has over the years raised increasing interest.

NK Cells

NK cells are large granular lymphocytes that are involved in the innate immune response primarily against invading pathogens and transformed cells [7]. They represent an important source of immunoregolatory cytokines and are capable of killing target cells through a direct mechanism or an antibody-dependent cellular cytotoxicity system [8]. Remarkably NK cells, that do not rearrange genes encoding for antigen receptors, are capable of killing cancer cells without the need of recognizing specific antigens, thus far poorly defined in the greater part of malignancies, or prior sensitization [9]. For these reasons, these cytotoxic cells are attractive effectors for the design of innovative immuno-based therapeutic strategies.

Human mature NK cells, which represent 10-15% of all peripheral blood lymphocytes, are defined by the presence on the cell surface of the CD56 antigen and the lack of CD3. Based on the density of surface expression of CD56 and the presence or absence of CD16, it is possible to distinguish two different populations of NK cells with different functional properties [10-13]. NK cells which express CD56 at low density and high levels of CD16, representing approximately 90% of the entire population, are natural cytotoxic effectors against sensitive targets [14], while CD56 bright NK cells, which do not express the CD16, are the primary population that produces immunoregolatory cytokines including interferon (IFN)γ and tumor-necrosis-factor (TNF)α [15].

NK cells are capable of recognizing molecules of the major histocompatibility complex (MHC), class-I or class I-like, which are expressed on target cells. Through these receptors, the cell is able to discriminate between autologous normal cells and foreign or transformed cells which have lost or down-modulated the self-MHC molecules expression; this represents the so-called "missing-self" hypothesis [16]. In addition to these antigens, each NK cell presents a variety of different activatory and inhibitory receptors, which allow the modulation of NK cell function [17-19]; the cytotoxic activity is ultimately regulated by the balance of these activatory and inhibitory signals.

Use of Cytokines

Many studies have reported a defective NK activity in cancer patients at diagnosis [20-25], while results for the remission phase of the disease are less conclusive [26-29]. Previous reports have shown that leukemic blasts may be susceptible to the lytic action of lymphokine-activated killer (LAK) cells [30,31]. Nevertheless, in many leukemic cases at presentation blasts are resistant to autologous killing [32], while LAK activity directed against autologous blasts may be generated from peripheral blood lymphocytes of leukemic patients in remission [33].

With the goal of optimizing the immunologic control of neoplasia, different cytokines have been tested in different *in vitro* and *in vivo* settings. So far, only few of them have reached clinical use.

Interkeukin (IL)-2 and IL-15 are structurally related cytokines and have demonstrated overlapping functions, as well as distinct roles, in inducing NK proliferation and enhancing cytotoxicity [34,35]. The anti-leukemic potential of IL-2 treatment *in vivo* has been suggested for acute myeloid leukemia (AML) patients with a limited proportion of residual marrow blasts [36-38]. It is well tolerated even in elderly patients and may be used according to different schedules and dosages [39,40]. The role of IL-2-activated NK cells has also been investigated in patients who have undergone an allogeneic bone marrow transplant for chronic myeloid leukemia (CML) [41]; in this setting, a clear correlation between the generation of lytic activity against host-derived CML targets and the risk of relapse has been established. Moreover, activated NK cells have been used to suppress primitive leukemic progenitors from CML patients in long-term autologous cultures, suggesting that autologous IL-2 activated NK cells with potent major histocompatibility complex unrestricted cytotoxic activity are capable of suppressing malignant hematopoiesis [42].

Different data generated *in vitro* and *in vivo* suggest that IL-15 may play an important role in anti-tumor activity: it induces the expression of mRNA for perforin and granzymes in murine lymphocytes [43] and prolongs tumor remission induced by cyclophosphamide in rhabdomyosarcoma-bearing mice, supposedly via the stimulation of subpopulations of NK cells [44].

IL-12 is structurally distinct, has a modest proliferative effect on NK cells, but is capable, alone, of enhancing cytotoxicity [45]. It is known as NK cell stimulation factor [46]; it enhances NK activity and specific cytotoxic T-cell (CTL) response [47], and induces NK and

T cells to produce IFNγ and TNFα [48]. These and other effects probably account for the ability of IL-12, alone or in combination with IL-2, to increase the lytic activity of peripheral blood mononuclear cells (PBMC) against tumor cell lines [49] and primary allogeneic [50] and autologous [51] leukemic blasts, to correct the defective cytotoxic activity of neoplastic patients at diagnosis [52] and to induce anti-neoplastic activity in murine cancer models [53].

Early growth factors, such as the ligands for the c-kit (KL) and the Flt-3 (FL) molecules, alone or in combination with other cytokines and/or drugs, are also under investigation with the aim of increasing NK cell expansion with promising results [54].

The Case of Haploidentical Stem Cell Transplantation

Recently, NK clones of donor origin have been established in the post-transplant period of haploidentical KYR mismatch SCT, when a donor-versus-recipient NK cell alloreactivity is present [55]; in this case, the inhibitory receptors expressed on donor NK cells do not recognize the MHC-class I molecules of the recipient and the cells can be activated by alternative activatory pathways. NK cells are the first lymphoid population to recover after an haploidentical SCT and may mediate both a GVL effect, by recognizing the tumor cells, and an anti-rejection activity through their activation against residual host T cell. The restriction of NK cell alloreactivity towards the hematopoietic compartment explains its lack of association with graft-versus-host disease (GVHD). Moreover, NK cells may recognize host dendritic cells implicated in the pathogenesis of GVHD, thus helping in the prevention of this phenomenon [56]. It has been demonstrated that alloreactive NK clones are capable of killing recipients' leukemic cells and that their presence is associated with a higher probability of overall survival and disease-free survival [57]. In addressing the GVL potential of donor-versus-recipient NK cell alloreactivity, it was found that 100% of AML, but only a minority of acute lymphoid leukemias (ALL), were killed by alloreactive NK clones. This may depend on the presence of specific adhesion molecules expressed on target cells capable of playing a decisive role on NK cell-mediated cytotoxicity. These finding have given a further input to the procedure of haploidentical SCT, which has now become a true alternative to matched unrelated SCT.

No reports are at the moment present in the literature regarding the infusion of *ex vivo* generated alloreactive NK clones. Nevertheless, this may represents a possible and very promising approach, in view of the possibility of preliminarily assessing the *in vitro* killing activity against the leukemic blast.

Generation and Expansion of Cytotoxic NK Cells

The aim of NK cell *ex vivo* generation and expansion is to increase the total number and to activate freshly isolated cytotoxic NK cells. We and others have demonstrated the possibility of expanding cytotoxic NK cells with killing activity against autologous and allogeneic blasts obtained from acute leukemia patients in complete remission (CR) [58-60]. Under the same culture conditions that enable the expansion of NK cells from normal PBMC

[61], a similar 40-fold expansion could be consistently achieved from the peripheral blood of both AML and ALL patients. This expanded population of effector cells presents an intact signal transduction apparatus and a preserved capacity to produce cytokines important in the cytolytic process. The cytotoxic activity against NK-susceptible and NK-resistant tumor cell lines was comparable to that observed with normal donors. A low basal cytotoxic activity against autologous blasts cryopreserved at diagnosis was also observed. In our attempt to increase this level of cytotoxicity, NK cells were incubated for 24 hours with different combinations of activating cytokines. IL-2 and IL-15 were able to maximally increase the cytotoxic activity against autologous blasts in all the samples tested, from both AML and ALL patients, including also the more resistant blasts.

Most of the reports present in the literature regarding patients affected by ALL have failed to generate *in vitro* cytotoxic effectors with lytic activity against autologous blasts. Great interest has therefore been generated, particularly in this setting, by all the procedures that may be applied with the aim of controlling MRD. These findings suggest a new possible immunotherapeutic strategy based on the expansion and infusion of cytotoxic NK cells, or on the cryopreservation of PBMC at the attainment of CR that could be subsequently expanded in the presence of persistent disease or early relapse.

The cytolytic properties of this expanded cell population have been recently further confirmed *in vivo* in a NOD/SCID mouse tumor model that showed a consistent reduction of AML load after adoptive transfer of autologous expanded IL-2- and IL-15-activated NK cells into tumor-transplanted mice [62].

Infusion of NK Cells and Good Manufactory Practice

Few reports, most in abstract form, are present in the literature regarding different approaches of *in vivo* infusion of NK cells, with somehow promising results. The number of treated patients is so far very small. NK cells have been generally obtained after leukapheresis followed by T-cell depletion and NK-cell enrichment. This has been carried out in the setting of haploidentical SCT in patiens with mixed chimerism or relapse [63-65], or after SCT from sibling or unrelated donors [66]. These reports demonstrate that the *ex-vivo* purification of NK cells from leukapheresis products is feasible and no important adverse events or onset of GVHD have been so far ascribed to the infusion [67].

It is of course mandatory that the generation, expansion and activation of cytotoxic cells is performed under "Good Manufactory Practice" (GMP) compliant procedures. In order to verify whether NK cells could be expanded under GMP conditions and whether these expanded effectors may potentially exert anti-leukemic activity, we produced a number of experiments under different culture conditions. When Ficoll density gradient separated PBMC, after NK cell enrichment, were co-cultured with autologous irradiated mononuclear feeder cells in addition to IL-2 and IL-15, a significant expansion of cytotoxic NK cells with killing activity against primary AML blasts has been achieved [68].

Conclusion

Acute leukemias are very aggressive family of diseases characterized by an overall poor long-term outcome, particularly in the adult age. The majority of the patients can be induced into CR following intensive chemotherapy protocols; nevertheless, most of them present early disease recurrence. Thanks to the modern diagnostic technologies, a strict and sharp monitoring of the disease is now possible for most cases. For these patients, times are mature for the design of vaccination programs based on the infusion of cytotoxic effectors and aimed at controlling or eradicating MRD. Clinical protocols, that should be based on *ex-vivo* expanded polyclonal autologous or allogeneic NK cells or on the infusion of *ex-vivo* produced NK clones, appear feasible. The experience so far developed suggests that the infusion of NK cells should induce very limited toxicity in the autologous setting and no or a very low risk of GVHD in the allogeneic situation, thus avoiding the potential complications associated to donor T-lymphocyte infusions. Moreover, cytokines such as IL-2 and IL-15 added to the culture medium and not directly infused to the patients as requested in other therapeutic protocols, should be devoid of the toxicities and side-effects that are associated with *in vivo* cytokine infusions. Phase I-II clinical trials are of course needed and awaited to verify the possible clinical role of these effectors in the control of hematologic malignancies.

References

[1]   Horowitz MM, Gale RP, Sondel PM, Goldman JM, Kersey J, Kolb HJ, Rimm AA, Ringden O, Rozman C, Speck B. Graft-versus-leukemia reactions after bone marrow transplantation. *Blood* 1990; 75: 555-562

[2]   Falkenburg JH, Smit WM, Willemze R. Cytotoxic T-lymphocyte (CTL) responses against acute or chronic myeloid leukemia. *Immunol Rev* 1997; 157: 223-230

[3]   Kolb HJ, Mittermuller J, Clemm C, Holler E, Ledderose G, Brehm G, Heim M, Wilmanns W. Donor leukocyte transfusions for treatment of recurrent chronic myelogenous leukemia in marrow transplant patients. *Blood* 1990; 76: 2462-2465

[4]   Kolb HJ, Schattenberg A, Goldman JM, Hertenstein B, Jacobsen N, Arcese W, Ljungman P, Ferrant A, Verdonck L, Niederwieser D. Graft-versus-leukemia effect of donor lymphocyte transfusions in marrow grafted patients. European Group for Blood and Marrow Transplantation Working Party Chronic Leukemia. *Blood* 1995; 86: 2041-2050

[5]   Collins RH Jr., Shpilberg O, Drobyski WR, Porter DL, Giralt S, Champlin R, Goodman SA, Wolff SN, Hu W, Verfaillie C, List A, Dalton W, Ognoskie N, Chetrit A, Antin JH, Nemunaitis J. Donor leukocyte infusions in 140 patients with relapsed malignancy after allogeneic bone marrow transplantation. *J Clin Oncol* 1997; 15: 433-444

[6]   Torelli GF, Orsini E, Guarini A, Kell J, Foa R. Developmental approaches in immunological control of acute myelogenous leukaemia. *Best Pract Res Clin Haematol* 2001; 14: 189-209

[7]   Robertson MJ, Ritz J. Biology and clinical relevance of human natural killer cells. *Blood* 1990; 76: 2421-2438

[8]   Trinchieri G. Biology of natural killer cells. *Adv Immunol* 1989; 47: 187–376

[9]   Allavena P, Damia G, Colombo T, Maggioni D, D'Incalci M, Mantovani A. Lymphokine-activated killer (LAK) and monocyte-mediated cytotoxicity on tumor cell lines resistant to antitumor agents. *Cell Immunol* 1989; 120: 250–258

[10]   Cooper MA, Fehniger TA, Caligiuri MA. The biology of human natural killer-cell subsets. *Trends in Immunology* 2001; 22: 633–640

[11]   Cooper MA, Fehniger TA, Turner SC, Chen KS, Ghaheri BA, Ghayur T, Carson WE, Caligiuri MA. Human natural killer cells: a unique innate immunoregulatory role for the CD56 (bright) subset. *Blood* 2001; 97: 3146–3151

[12]   Caligiuri MA, Zmuidzinas A, Manley TJ, Levine H, Smith KA, Ritz J. Functional consequences of interleukin 2 receptor expression on resting human lymphocytes. Identification of a novel natural killer cell subset with high affinity receptors. *J Exp Med* 1990; 171: 1509–1526

[13]   Baume DM, Robertson MJ, Levine H, Manley TJ, Schow PW, Ritz J. Differential responses to interleukin 2 define functionally distinct subsets of human natural killer cells. *Eur J Immunol* 1992; 22: 1–6

[14]   Nagler A, Lanier LL, Cwirla S, Phillips JH. Comparative studies of human FcRIII-positive and negative natural killer cells. J Immunol 1989; 143: 3183–3191

[15]   Perussia B, Chen Y, Loza MJ. Peripheral NK cell phenotypes: multiple changing of faces of an adapting, developing cell. *Mol Immunol* 2005; 42: 385-395

[16]   Ljunggren HG, Karre K. In search of the 'missing self': MHC molecules and NK cell recognition. Immunol Today 1990; 11: 237-244

[17]   Lanier LL. NK cell receptors. *Annu Rev Immunol* 1998; 16: 359-393

[18]   Moretta L, Moretta A. Killer immunoglobulin-like receptors. *Curr Opin Immunol* 2004; 16: 626-633

[19]   Lanier LL. NK cell recognition. *Annu Rev Immunol* 2005; 23: 225-274

[20]   Tursz T, Dokhelar MC, Lipinski M, Amiel JL. Low natural killer cell activity in patients with malignant lymphoma. *Cancer* 1982; 50: 2333-2335

[21]   Ruco LP, Procopio A, Maccallini V, Calogero A, Uccini S, Annino L, Mandelli F, Baroni CD. Severe deficiency of natural killer activity in the peripheral blood of patients with hairy cell leukemia. *Blood* 1983; 61: 1132-1137

[22]   Frydecka I. Natural killer cell activity during the course of disease in patients with Hodgkin's disease. *Cancer* 1985; 56: 2799-2803

[23]   Nakagomi H, Petersson M, Magnusson I, Juhlin C, Matsuda M, Mellstedt H, Taupin JL, Vivier E, Anderson P, Kiessling R. Decreased expression of the signal-transducing ζ chains in tumor-infiltrating T-cells and NK cells of patients with colorectal carcinoma. *Cancer Res* 1993; 53: 5610-5612

[24]   Lai P, Rabinowich H, Crowley-Nowich PA, Bell MC, Mantovani G, Whiteside TL. Alterations in expression and function of signal-transducing proteins in tumor-associated T and natural killer cells in patients with ovarian carcinoma. *Clin Cancer Res* 1996; 2: 161-173

[25]   Kono K, Ressing ME, Brandt RMP, Melief CJ, Potkul RK, Andersson B, Petersson M, Kast WM, Kiessling R. Decreased expression of signal-transducing ζ chain in peripheral T cells and natural killer cells in patients with cervical cancer. *Clin Cancer Res* 1996; 2: 1825-1828

[26]   Yoda Y, Abe T, Tashiro A, Hirosawa S, Kawada K, Onozawa Y, Adachi Y, Shishido
        H, Nomura T. Normalised natural killer (NK) cell activity in long-term remission of
        acute leukaemia. *Br J Haematol* 1983; 55: 305-309

[27]   Dickinson AM, Proctor SJ, Jacobs E, Reid MM, Walker W, Craft AW, Kernahan J.
        Natural killer cell activity in childhood acute lymphoblastic leukaemia in remission.
        Br J Haematol 1985; 59: 45-53

[28]   Pizzolo G, Trentin L, Vinante F, Agostini C, Zambello R, Masciarelli M, Feruglio C,
        Dazzi F, Todeschini G, Chilosi M, *et al*. Natural killer cell function and lymphoid
        subpopulations in acute non-lymphoblastic leukaemia in complete remission. *Br J
        Cancer* 1988; 58: 368-372

[29]   Trentin L, Pizzolo G, Feruglio C, Zambello R, Masciarelli M, Bulian P, Agostini C,
        Vinante F, Zanotti R, Semenzato G. Functional analysis of cytotoxic cells in patients
        with acute nonlymphoblastic leukemia in complete remission. *Cancer* 1989; 64: 667-
        672

[30]   Adler A, Chervenich PA, Whiteside TL, Lotzova E, Herberman RB. Interleukin 2
        induction of lymphokine-activated killer (LAK) activity in the peripheral blood and
        bone marrow of acute leukemia patients. Feasibility of LAK generation in adult
        patients with active disease and in remission. *Blood* 1988; 71: 709-716

[31]   Fierro MT, Liao XS, Lusso P, Bonferroni M, Matera L, Cesano A, Lista P, Arione R,
        Forni G, Foa R. In vitro and in vivo susceptibility of human leukemic cells to
        lymphokine activated killer activity. *Leukemia* 1988; 2: 50-54

[32]   Foa R, Meloni G, Tosti S, Novarino A, Fenu S, Gavosto F, Mandelli F. Treatment of
        acute myeloid leukaemia patients with recombinant interleukin 2: a pilot study. *Br J
        Haematol* 1991; 77: 491-496

[33]   Lauria F, Raspadori D, Rondelli D, Ventura MA, Foa R. In vitro susceptibility of
        acute leukemia cells to the cytotoxic activity of allogeneic and autologous lymphokine
        activated killer (LAK) effectors: correlation with the rate and duration of complete
        remission and with survival. *Leukemia* 1994; 8: 724-728

[34]   Waldmann T, Tagaya Y, Bamford R. Interleukin-2, interleukin-15, and their
        receptors. *Int Rev Immunol* 1998; 16: 205-226

[35]   Fehniger TA, Caligiuri MA. Interleukin 15: biology and relevance to human disease.
        Blood 2001; 97: 14-32

[36]   Foa R, Meloni G, Tosti S, Novarino A, Fenu S, Gavosto F, Mandelli F. Treatment of
        acute myeloid leukaemia patients with recombinant interleukin 2: a pilot study. *Br J
        Haematol* 1991; 77: 491-496

[37]   Meloni G, Foa R, Vignetti M, Guarini A, Fenu S, Tosti S, Tos AG, Mandelli F.
        Interleukin-2 may induce prolonged remissions in advanced acute myelogenous
        leukemia. *Blood* 1994; 84: 2158-2163

[38]   Meloni G, Trisolini SM, Capria S, Torelli GF, Baldacci E, Torromeo C, Valesini G,
        Mandelli F. How long can we give interleukin-2? Clinical and immunological
        evaluation of AML patients after 10 or more years of IL2 administration. *Leukemia*
        2002; 16: 2016-2018

[39]   Caligiuri MA, Murray C, Robertson MJ, et al. Selective modulation of human natural
        killer cells in vivo after prolonged infusion of low dose recombinant interleukin 2. *J
        Clin Invest* 1993; 91: 123-132

[40] Farag SS, George SL, Lee EJ, Baer M, Dodge RK, Becknell B, Fehniger TA, Silverman LR, Crawford J, Bloomfield CD, Larson RA, Schiffer CA, Caligiuri MA. Postremission therapy with low-dose interleukin 2 with or without intermediate pulse dose interleukin 2 therapy is well tolerated in elderly patients with acute myeloid leukemia: Cancer and Leukemia Group B study 9420. *Clin Cancer Res* 2002; 8: 2812–2819

[41] Hauch M, Gazzola MV, Small T, Bordignon C, Barnett L, Cunningham I, Castro-Malaspinia H, O'Reilly RJ, Keever CA. Anti-leukemia potential of interleukin-2 activated natural killer cells after bone marrow transplantation for chronic myelogenous leukemia. *Blood* 1990; 75: 2250-2262

[42] Cervantes F, Pierson BA, McGlave PB, Verfaillie CM, Miller JS. Autologous activated natural killer cells suppress primitive chronic myelogenous leukemia progenitors in long-term culture. *Blood* 1996; 87: 2476-2485

[43] Ye W, Young JD, Liu CC. Interleukin-15 induces the expression of mRNAs of cytolytic mediators and augments cytotoxic activities in primary murine lymphocytes. *Cell Immunol* 1996; 174: 54-62

[44] Evans R, Fuller JA, Christianson G, Krupke DM, Troutt AB. IL-15 mediates anti-tumor effects after cyclophosphamide injection of tumor-bearing mice and enhances adoptive immunotherapy: the potential role of NK cell subpopulations. *Cell Immunol* 1997; 179: 66-73

[45] Gately MK, Renzetti LM, Magram J, Stern AS, Adorini L, Gubler U, Presky DH. The interleukin-12/interleukin-12-receptor system: role in normal and pathologic immune responses. *Annu Rev Immunol* 1998; 16: 495-521

[46] Kobayashi M, Fitz L, Ryan M, Hewick RM, Clark SC, Chan S, Loudon R, Sherman F, Perussia B, Trinchieri G. Identification and purification of natural killer cell stimulatory factor (NKSF), a cytokine with multiple biologic effects on human lymphocytes. *J Exp Med* 1989; 170: 827-845

[47] Chehimi J, Valiante NM, D'Andrea A, Rengaraju M, Rosado Z, Kobayashi M, Perussia B, Wolf SF, Starr SE, Trinchieri G. Enhancing effect of natural killer cell stimulatory factor (NKSF/interleukin-12) on cell-mediated cytotoxicity against tumor-derived and virus-infected cells. *Eur J Immunol* 1993; 23: 1826-1830

[48] Chan SH, Perussia B, Gupta JW, Kobayashi M, Pospisil M, Young HA, Wolf SF, Young D, Clark SC, Trinchieri G. Induction of interferon gamma production by natural killer cell stimulatory factor: characterization of the responder cells and synergy with other inducers. *J Exp Med* 1991; 173: 869-879

[49] Rossi AR, Pericle F, Rashleigh S, Janiec J, Djeu JY. Lysis of neuroblastoma cell lines by human natural killer cells activated by interleukin-2 and interleukin-12. *Blood* 1994; 83: 1323-1328

[50] Uharek L, Zeis M, Glass B, Steinmann J, Dreger P, Gassmann W, Schmitz N, Muller-Ruchholtz W. High lytic activity against human leukemia cells after activation of allogeneic NK cells by IL-12 and IL-2. *Leukemia* 1996; 10: 1758-1764

[51] Vitale A, Guarini A, Latagliata R, Cignetti A, Foa R. Cytotoxic effectors activated by low-dose IL-2 plus IL-12 lyse IL-2-resistant autologous acute myeloid leukaemia blasts. *Br J Haematol* 1998; 101: 150-157

[52]   Soiffer RJ, Robertson MJ, Murray C, Cochran K, Ritz J. Interleukin-12 augments cytolytic activity of peripheral blood lymphocytes from patients with hematologic and solid malignancies. *Blood* 1993; 82: 2790-2796

[53]   Wigginton JM, Komschlies KL, Back TC, Franco JL, Brunda MJ, Wiltrout RH. Administration of interleukin 12 with pulse interleukin 2 and the rapid and complete eradication of murine renal carcinoma. *J Natl Cancer Inst* 1996; 88: 38-43

[54]   Farag SS, Caligiuri MA. Human natural killer cell development and biology. *Blood Reviews* 2006: 20: 123-137

[55]   Ruggeri L, Capanni M, Casucci M, Volpi I, Tosti A, Perruccio K, Urbani E, Negrin RS, Martelli MF, Velardi A. Role of natural killer cell alloreactivity in HLA-mismatched hematopoietic stem cell transplantation. *Blood* 1999; 94: 333-339

[56]   Ruggeri L, Capanni M, Urbani E, Perruccio K, Shlomchik WD, Tosti A, Posati S, Rogaia D, Frassoni F, Aversa F, Martelli MF, Velardi A. Effectiveness of donor natural killer cell alloreactivity in mismatched hematopoietic transplants. *Science* 2002;   295
2097-2100

[57]   Giebel S, Locatelli F, Lamparelli T, Velardi A, Davies S, Frumento G, Maccario R, Bonetti F, Wojnar J, Martinetti M, Frassoni F, Giorgiani G, Bacigalupo A, Holowiecki J. Survival advantage with KIR ligand incompatibility in hematopoietic stem cell transplantation from unrelated donors. *Blood* 2003; 102: 814-819

[58]   Torelli GF, Guarini A, Palmieri G, Breccia M, Vitale A, Santoni A, Foa R. Expansion of cytotoxic effectors with lytic activity against autologous blasts from acute myeloid leukaemia patients in complete haematological remission. *Br J Haematol* 2002; 116: 299-307

[59]   Guven H, Gilljam M, Chambers BJ, Ljunggren HG, Christensson B, Kimby E, Dilber MS. Expansion of natural killer (NK) and natural killer-like T (NKT)-cell populations derived from patients with B-chronic lymphocytic leukemia (B-CLL): a potential source for cellular immunotherapy. *Leukemia* 2003; 17: 1973-1980

[60]   Torelli GF, Guarini A, Maggio R, Alfieri C, Vitale A, Foa R. Expansion of natural killer cells with lytic activity against autologous blasts from adult and pediatric acute lymphoid leukemia patients in complete hematologic remission. *Haematologica/THJ* 2005; 90: 785-92

[61]   Perussia B, Ramoni C, Anegon I, Cuturi MC, Faust J, Trinchieri G. Preferential proliferation of natural killer cells among peripheral blood mononuclear cells cocultured with B lymphoblastoid cell lines. *Nat Immun Cell Growth Regul* 1987; 6: 171-188

[62]   Siegler U, Kalberer CP, Nowbakht P, Sendelov S, Meyer-Monard S, Wodnar-Filipowicz A. Activated natural killer cells from patients with acute myeloid leukemia are cytotoxic against autologous leukemic blasts in NOD/SCID mice. *Leukemia.* 2005; 19: 2215-22

[63]   Passweg JR, Tichelli A, Meyer-Monard S, Heim D, Stern M, Kuhne T, Favre G, Gratwohl A. Purified donor NK-lymphocyte infusion to consolidate engraftment after haploidentical stem cell transplantation. *Leukemia* 2004; 18: 1835-1838

[64]   Gentilini C, Hilbers U, Hartung G, Lange T, Kliem C, Hegenbart U, Zeis M, Huppert V, Glass B, Uhrberg M, Schmitz N, Thiel E, Niederwieser D, Uharek L. Early transfer of highly purified alloreactive CD56+CD3- NK cells after haploidentical stem

cell transplantation: promising results of a phase I study. *Bone Marrow Transplantation* 2006; 37: S35, O257

[65] Koehl U, Sorensen J, Esser R, Becker M, Huenecke S, Munkelt D, Tonn T, Seifried E, Klingebiel T, Bader P, Passweg J, Schwabe D. Immunotherapy with highly purified CD56+CD3- natural killer cells for paediatric patients after haploidentical stem cell transplantation. *Bone Marrow Transplantation* 2006; 37: S35, O258

[66] Slavin S, Morecki S, Shapira M, Slavin S, Morecki S, Shapira M, Samuel S, Ackerstein A, Gelfand Y, Resnick I, Bitan M, Or R. Immunotherapy using rIL-2 activated mismatched donor lymphocytes positively selected for CD56+ for the treatment of resistant haematologic malignancies after stem cell transplantation. *Bone Marrow Transplantation* 2004; 33: S174, P686

[67] Passweg JR, Koehl U, Uharek L, Meyer-Monard S, Tichelli A. Natural-killer-cell-based treatment in haematopoietic stem-cell transplantation. *Best Pract Res Clin Haematol* 2006; 19: 811-824

[68] Torelli GF, Mascolo MG, Ricciardi MR, De Propris MS, Milano F, Lucarelli B, Malandruccolo L, Iannella E, Iori AP, Tafuri A, Guarini A, Foà R. Expansion of natural killer cells with lytic activity against acute myeloid leukemia blasts under good manufacturing practice conditions. *European Hematology Association*, Wien 2007, Abs n. 0593.

In: Natural Killer T-Cells: Roles, Interactions and Interventions   ISBN: 978-1-60456-287-3
Editor: Nathan V. Fournier, pp. 155-171   © 2008 Nova Science Publishers, Inc.

Chapter 7

Natural Killer Cell Receptor NKG2A/HLA-E Interaction Dependent Differential Thymopoiesis of Hematopoietic Progenitor Cells Influences the Outcome of HIV Infection

*Edmond J. Yunis[*][a], Viviana Romero[a], Felipe Diaz-Giffero[a],
Joaquin Zuñiga[a, b] and Prasad Koka[b]*

[a]Department of Cancer Immunology and AIDS, Dana-Farber Cancer Institute and
Department of Pathology, Harvard Medical School, Boston MA
[b] Laboratory of Immunobiology and Genetics, Instituto Nacional de Enfermedades
Respiratorias, Mexico D.F. , Mexico
[c] Laboratory of Stem Cell Biology, Torrey Pines Institute for Molecular Studies,
San Diego, CA 92121, USA

Abstract

HIV infection and its outcome is complex because there is great heterogeneity not only in clinical presentation, incomplete clinical information of markers of immunodeficiency and in measurements of viral loads. Also, there many gene variants that control not only viral replication but immune responses to the virus; it has been difficult to study the role of the many AIDS restricting genes (ARGs) because their influence vary depending on the ethnicity of the populations studies and because the cost to follow infected individuals for many years. Nevertheless, at least genes of the major histocompatibility locus (MHC) such as HLA alleles have been informative to classify

[*] **Correspondence:** Edmond J. Yunis, M.D., Department of Cancer Immunology and AIDS, Dana Farber Cancer Institute, Harvard, Medical School, 44 Binney Street, Boston, MA. 02115. Phone: (617) 632 3347 Fax: (617) 632 5151. Email: edmond_yunis@dfci.harvard.edu

infected individuals following HIV infection; progression to AIDS and long-term-non-progressors (LTNP). For example, progressors could be defined as up to 5 years, up to 11 years or as we describe in this report up to 15 years from infection, and LTNP could be individuals with normal CD4+ T cell counts for more than 15 years with or without high viral loads. In this review, we emphasized that in the studies of ARGs the HLA alleles are important in LTNP; HLA-B alleles influencing the advantage to pathogens to produce immune defense mediated by CD8+ T cells (cognate immunuity). Our main point we make in this report is that contrary to recent reports claiming that this dominant effect was unlikely due to differences in NK activation through ligands such as HLA-Bw4 motif, we believe that cognate immunity as well as innate inmmunity conferred by NK cells are involved. The main problem is that HLA-Bw4 alleles can be classified according the aminoacid in position 80. Isoleucine determines LTNP, which is a ligand for 3DS1. Such alleles did not include HLA-B*44, B*13 and B*27 which have threonine at that position. The authors have not considered the fact that in addition to the NK immunoglobulin receptors, NK receptors can be of the lectin like such as NKG2A/HLA-E to influence the HIV infection outcome. HLA-Bw4 as well as HLA-Bw6 alleles can be classified into those with threonine or methionine in the second position of their leader peptides. These leader peptides are ligands for NKG2A in which methionine influences the inhibitory role of NKG2A for killing infected targets. Functional studies have not been done as well as studies of these receptors in infected individuals. However, analyses of the leader peptides of HLA-B alleles in published reports, suggested that threonine in the second position can explain the importance of HLA-B*57, B*13, B*44 as well as certain Bw6 alleles in LNTP. In addition, we analyzed the San Francisco database that was reported and found that the association of HLA-B alleles with LNTP or with progressors can be due to the presence of threonine or methionine in their second position. Therefore, studies of outcome of HIV infection should include not only mechanisms of cognate immunity mediated by peptides and CD8+ T cells but also, NK receptors of two types, NKG2A as well as 3DS1. We propose that the SCID mouse should be used to understand mechanisms mediated by many of the ARGs especially the importance of thymus derived cells as well as NK receptor interactions with their ligands in this experimental animal transplanted with human stem cells, thymus or NK cells obtained from individuals of known HLA genotypes.

I. Introduction

AIDS is not generally considered a genetic disease because there is a great heterogeneity of clinical presentation in part determined by gene variants that control to same extent virus replication and immunity [1-3]. Noteworthy of mentioning is the fact that there is not much information about genes that protect individuals exposed to the virus and do not get infected. Also, it is important to mention that those that get infected develop pathology at different times following the infection, for example those who died from AIDS at 1-5 years from infection, but the definition of such fast progressors should include those where the immunodeficiency and progression to AIDS could go up to 10-11 years or more from the time of the infection. More difficult perhaps, is to define long-term infection since there are patients that progress to immunodeficiency 15-20 or more years after infection and a group of the long-term non progressors that maintain normal CD4+ T cell counts and absence of viremia for more than 20 years without treatment. The heterogeneity of clinical presentation cannot be completely explained, but it is possibly related to the large number of genetic and non-genetic contributing factors: innate, humoral and cell mediated immune responses [4] and the AIDS restricting genes (ARGs). In addition, longitudinal studies have an additional difficulty because some patients are treated with antiretroviral drugs to prevent or diminish the possible progression to AIDS.

Our review will attempt to emphasize that studies of CD8+ T cell immunity should be performed in conjunction with NK receptor/ligand interactions and their variants. We will summarize evidence that HIV infection involves several immune mechanisms resulting in long term non progression or progression to AIDS in which intrinsic, innate and cognate immunity are important. However, genetic associations with protection from or progression to disease are complex and could vary, related to the ethnicity of the populations studied [5].

II. Immune Functions in HIV Infection

A. Cognate Immunity

Cognate immunity has been extensively studied involving primarily CD4+ and CD8+ T cells interacting with MHC alleles presenting HIV-1 peptides on antigen presenting cells [6-12]. Such cells may mediate protection by production of cytotoxic cells or cytokines such as IFN gamma and IL-12 [13]. Macrophages [14], DC and plasmocytoid DCs are type 1 IFN secretion inducers of IL-12. However, several viruses escape MHC class I restricted cytotoxic T lymphocytes (CTL) responses by down-regulating their expression on infected cell surface [15]. In this regard, protection from infection may have genetic basis but it varies in different populations [5]; these associations are complex and difficult to assess because in most cases is difficult to know the date of incidence of infection, the date of presence of viremia, or the date of decrease of CD4 counts. In many cases the studies could be difficult to reproduce because of genetic stratification and population size of case-control studies. MHC studies, in many cases, are incomplete and without strict criteria for population sizes and epidemiological parameters that explain the lack of consistencies of the alleles associated with disease progression [16]. Also, some of the MHC associations maybe present because of genes that are in linkage disequilibrium with HLA alleles, for example TNF [17], that are involved in the regulation of level of HIV-1 viremia since it causes increased viral replication in infected monocytes. This occurs through the activation of the transcription factor NFkB, which binds to the HIV-1 long terminal repeat, causing increased levels of HIV-1 transcription [18-20].

HLA Alleles, and Clinical Outcome of HIV-1 Infection

HLA alleles of class I (A, B and C) and II (DR, DQ and DP) have a large variation between individuals and populations, which could provide recognition of virus agents to which they have been exposed [4]. Since HIV infects immune cells to produce proliferation, spread and CD4+ T lymphocyte damage, the HLA alleles could influence the time from infection to AIDS progression [21]. For example, HLA-Cw4, a ligand for KIR 2DL1 influences the time to develop AIDS, [22]. These findings could have been due to genes in random association with HLA-B and C alleles such as TNF [17]. The contribution of HLA-B*35 alleles has also been postulated [23]. More importantly, HLA influence was observed in relation to HIV-1 subtypes where the HLA recognition motifs are associated to AIDS survival. Two HLA alleles, B*27 and B*57, have been reproducibly associated with long-term non-progression of AIDS [3, 24]. A controversial subject is the role of zygosity of HLA-A alleles in the progression to AIDS. For example HLA class I homozygotes for several alleles progress to AIDS faster than those heterozygous [23, 24]. Further, groups of alleles or supertypes grouped according the B pocket showed association with clinical outcome [16, 25]. The role of zygosity for HLA-Bw6 and HLA-Bw4 has been controversial, Bw6 was found associated with AIDS in Caucasians [25, 26] and Chinese [27] while the Bw4/Bw4 was associated with LTNP in Caucasians but not in Africans [28]. Therefore, genetic markers, ARGs result from different evolutionary effects that would explain that ethnicity and protection from AIDS and genetic effects on clinical outcome vary among populations [4, 5].

B. Innate Immunity

This type of immunity is mediated by a newly discovered endogenously expressed proteins that provide defenses against retroviral infection such as HIV-1 and murine leukemia virus (MLV) [29-31]. These proteins probably prevent the entry of HIV-1 in to cells by interaction with CD4, Natural Killer (NK) cells, monocytes and dendritic cells. In addition, several molecules secreted by these cells can protect against such infection, for example IL-12, chemokine receptors CCR4 and CCR5. In this regard, most HIV strains use the CCR5 as a co-receptor and thereby are sensitive to inhibition by the ligands of this receptor; also deletion of CCR5 prevents viral entry into cells [2]. HIV-1 infection is potently blocked in rhesus macaques by the cytoplamic component protein tripartite motif 5 (TRIM5). Interestingly, the TRIM5 protein in rhesus macaque is 87% identical to its human homolog. However, human TRIM5 proteins do not restrict HIV-1 [31]. Remarkably, one amino acid change in the protein sequence of human TRIM5 leads to full activity against HIV-1 infection [32, 33]. Even though, several SNPs for human TRIM5 have been identified, none of them corresponded to this specific residue. Differently, the endogenously expressed human cytidine deaminase APOBEC3G potently blocks HIV-1 infection [32]. However, the HIV-1 accessory protein vif is able to overcome this restriction allowing productive infection. Finally, the endogenously expressed mouse protein Fv1 potently block infection of MLV [29]. Overall, this demonstrates that during evolution, humans developed series of specie-specific blocks for retroviral infection.

In an extensive genetic epidemiological study, SNPs of TRIM5 were identified. In African Americans four alleles exhibited different frequencies in HIV-1 infected and uninfected individuals. SNP2 in the non-coding exon 1 and SNP3 in intron 1 were associated with increased risk of infection. [34]. These finding suggested that any modification of infection susceptibility afforded by particular TRIM5 alleles may be restricted to particular populations or types of exposure. There are population specific effects for AIDS modifying genes [33-36]. For example, the role of TRIM5 polymorphism in Europeans, the TRIM haplotype containing 136Q exhibited increased frequency among HIV-1 infected subjects compared with seronegative exposed individuals [34], which contrasts with the opposite results reported in African Americans [35, 36]. These findings suggest that either the genetic background, non-random association with other genes or the type of exposure may influence susceptibility to HIV-1 infection. However, TRIM5 variants do not influence disease progression. As it will be pointed out below, there are many genes that limit AIDS and such influence may be present in some ethnicities but not in all. This problem stresses the fact that research is needed to determine their frequencies in several ethnicities in order to include sufficient individuals in the calculations of statistical power.

According to estimations there are more than 90% of the genetic and non-genetic influences on AIDS progression that are still undiscovered. Less known is the estimation of these factors influencing genetic susceptibility to be infected by HIV-1. It was calculated that 21.1% of individuals infected who did not develop AIDS for 11 more years did so because they carried one of more protective ARGs. Among those that develop AIDS rapidly, within 5.5 years of HIV-1 infection, the single ARG effects were modest (2.4-8.1%) but the cumulative ARGs associated with rapid progression is high (40.9%). However, the estimates are approximately 10% for both groups including long-term non-progressors and progressors to AIDS. Available databases, although informative, possess problems for analyses, because during the patients follow up some of them are recruited after they have been treated with anti-retroviral drugs during the first 10-11 years from the time of infection. Thus the definition of immunodeficiency is based on CD4+ T cell counts only without consideration of viral loads. Also some patients without treatment maintain normal CD4+ T cell counts beyond 15 years from infection while they may have high viral loads.

NK cells

Studies of HIV-1 uninfected persons showed enhanced activity of NK functions despite many years of high-risk exposure demonstrating the importance of NK cells in immunity against HIV-1 infection [37]. NK cells functions involve receptors that interact with HLA ligands. HLA alleles are therefore important in disease progression because target infected cells are more susceptible to NK killing. For example, *nef*, a product of HIV-1 is known to diminish the levels of HLA-A and HLA-B expression of infected cells [38], whereas HLA-C, which is one ligand for NK receptors, is poorly expressed naturally [39]. By contrast, *nef* increases the level of HLA-E on the surface of infected cells [40]. Also, HLA-A expression is higher than HLA-B [39] and HLA-B expression is more inducible by IFN alpha than the HLA-A [41]. Therefore, in the environment of the HIV infection there would be more contribution of HLA-B alleles or leader peptides from HLA-B as well as that of HLA-E with HLA-B leader peptides to function in their interaction with NK receptors.

There are two different kinds of NK receptors, immunoglobulin-like such as the inhibitory and activating receptors that include the KIRs and also lectin-like such as the NKG2 receptors. The NKG2A is inhibitory and has as a ligand HLA-E and the NKG2D is an activating-type receptor and has as a ligand the MICA alleles for function, killing of infected cells or cancer cells [42, 43]. Also, some KIRs are expressed on a subset of NK cells with a memory phenotype [44] that suggests that they may regulate T cell as well as NK cell activity. Masking of inhibitory NK receptors on CTLs from HIV infected individuals by monoclonal antibody has produced increases of HIV-1 specific CTL activity [12] suggesting possible involvement of NK cells.

As proposed in this report, HLA-E and its interaction with NKG2A needs to be investigated in both high risk persons exposed to HIV that have or do not have infection or their role in the clinical outcome following HIV-1 infection. However, the MICA alleles, which are ligands for the lectin activating receptor NKG2D, were not associated with HIV-1 clinical progression [17].

It is well known that HLA-E folds to function, depending on the leader peptides from HLA-class I alleles (A, C and B). The critical aminoacid is methionine in second position, whereas threonine is accompanied by poor folding and defective function [45]. The leader peptides with methionine in the second position (VMAPKTVLL and VMAPRTLL) induce higher levels of HLA-E expression than those with threonine at Position 2. Also, they exhibited high affinity for soluble CD4 NKG2A molecules [45-47]. Despite the high expression of HLA-E due to nef, the presence of threonine in the leader peptides of HLA-B alleles render HLA-E to be poorly expressed [40].

HLA-B Ligands for NK Receptors: HLA-Bw4 and HLA-Bw6 Supertypes and the Leader Peptide of HLA-B alleles as Ligands for the NKG2A Receptor

The HLA-B alleles are the most important in HIV viral progression because they restrict infection via CD4/CD8 [48] and they are ligands for NK receptors, KIRs (3DL1 and 3DS1) [26] and lectin receptor (NKG2A) as we propose in this article.

HLA-Bw4 supertype comprising approximately 40% of HLA-B alleles is the ligand for NK receptors encoded by the Killer immunoglobulin receptor (KIR) gene complex on chromosome 19 [49]. This interaction could be due to loss of inhibition of the inhibitory receptor, 3DL1 in effector cells that causes the function of the activating receptor 3DS1. This was found experimentally, that the combination of homozygosity of HLA-Bw4 and KIR3DS1 epistasis influence AIDS progression [23]. Of interest, this interaction did not include the role of B*27 and B*44 in LNTP. The authors suggested that those alleles are involved in a different mechanism of protection from progression. In this report we are proposing a mechanism that needs to be investigated.

Table 1. The HLA Bw4 and Bw6 motifs and alleles

HLA Supertype		Amino-acid position				Corresponding alleles
		80	81	82	83	
Bw4	Ile80	Ile	Ala	Leo	Arg	B*15(13,16,17.23,24) B*2702, B*3601, B*4901, B*51, B*52.01 B*5301, B*5302, B*57, B*58, B*59
	Thr80	Thr	Ala	Leu	Arg	B*13 (01, 02, 04), B*3802, B*44 (02, 03, 04, 05, 08, 21)
		Thr	Leu	Leu	Arg	B*2705, B*2709, B*3701, B*4701
Bw6	Asn 80	Asp	Leu	Arg	Gly	B*07 (02, 03, 04, 05, 08, 09, 10, 14) B*08, B*14. B15*(01, 03, 04, 09, 15, 18, 22, 37
						45, 48, 13, 16, 17, 23, 24)) B*18, B*35, B*39, B*40, B*41, B*42, B*4501, B*4601, B*48, B*50, B*5401, B*55, B*56, B*67, B*73, B*7801, B*8101, B*82

In bold are alleles whose leader peptide has methionine in the second position and the rest have threonine in the second position [50].

All HLA-A and HLA-C alleles encode HLA-E binding peptides with methionine at second position [50]. Importantly, HLA-C alleles are poorly expressed on the surface of cells [39] and HLA-B has higher basal level of expression than HLA-A, and by far more inducible by IFN gamma and alpha than HLA-A gene [41]. Furthermore, HLA-B alleles are the dominant influence to mediate a possible co-evolution of HIV and HLA [44], that we believe should also include HLA-E loaded with class I peptides.

These findings are based on the fact that HLA-B alleles can be classified according to the presence of thr or met at P2 of the leader peptide [41, 43]. Almost all HLA-Bw4 alleles with the exception of HLA-B*38 encode leader peptides with Thr at P2 and HLA-Bw6 are divided into two groups, those encoding Met at P2 and those encoding Thr at P2 [41, 43, 51] (Table 2).

Table 2. Comparison of Bw4 and Bw6 frequencies based on the presence of methionine or threonine in second position of HLA-B leader peptides

aa 2nd position	Caucasians	African Americans	Asian Americans
Met	0.30	0.20	0.18
Thr	0.70	0.80	0.82
Met/Met	0.09	0.04	0.04
Met/Thr	0.42	0.32	0.27
Thr/Thr	0.49	0.64	0.69
Supertype alleles			
Bw6	0.61	0.74	0.55
Bw4	0.39	0.26	0.45
Supertype genotypes			
Bw6/Bw6	0.37	0.55	0.30
Bw4/Bw4	0.15	0.07	0.20
Bw6/Bw4	0.49	0.38	0.50

Statistical Analysis of Influence of Threonine in Second Position in Long Term Non Progressors (LTNP)

Re-analyses of the data published before (Flores et al) (table 3) the number of individuals with two copies of Threonine in the group of controllers was significantly higher than in the progressors. 14/20. (0.70) versus 3/19 (0.15) of the non-controllers p: 0.0006 OR: 12.44. Of interest, the frequency of individuals with TT uninfected was 42/108 (0.39). p: 0.010 OR:3.67.

Analyses of the San Francisco cohort reported before also demonstrated that Comparison of LTNP with progressors, either 0-10 years or 0-15 years showed that the presence of TT was significaltly higher in the LTNP than in 0-10 years progressors, 17/22 versus 11/44, p= 0.000005, OR=10.20 and between all progressors (0 to 15 years, 17/22 versus 20/76, p=0.00001, OR= 9.52. Likewise, the frequency of TT was higher in LTNP than in uninfected controls 20/76 versus 130/265, p =0.0004 OR= 2.70. These results should be confirmed using Kaplan-Meier plots to assess more accurately the importance of these findings using outcome of HIV viral infection and need to be studied in relation to the expression of NKG2A receptors on NK cells.

Table 3. Frequencies of leader peptides (two copies) of methionine (M) or threonine (T), influence in HIV progression

aa 2nd position	Intermediate Progressors (10-15 years)	Progressors (0-10 years)	LTNP	Non-infected controls
MM	7/32 (21.8)	14/44 (31.8)	1/22 (4.5)	24/265 (9.0)
MT	16/32 (50.0)	19/44 (43.2)	4/22 (18.2)	138/265 (52.0)
TT	9/32 (28.1)	11/44 (25.0)	17/22 (77.3)	130/265 (49.0)

III. Stem Cell Microenvironment in HIV Infection

HIV infection of stem cell microenvironments or niches causes hematopoietic inhibition and hence cytopenias [52-57]. Hematopoietic CD34+ progenitor stem cells are reported to be resistant to HIV-1 infection, *in vitro*, or *in vivo* [58, 59]. Those cells that experienced the indirect effects of HIV-1 infection exhibit inhibition of their multilineage hematopoiesis as determined by colony forming activity ex vivo [58, 60-62]. It is reported that the hematopoietic stem cell microenvironment is damaged due to the indirect effects of HIV-1 infection of the thymocytes on the CD34+ progenitor stem cells but in a reversible manner, in the human fetal Thymus/Liver conjoint hematopoietic organ of the transplanted chimeric severe combined immunodeficiency mouse (SCID-hu) model system [60, 62]. It is therefore highly plausible that this implanted human organ in the SCID-hu mouse, which serves as a niche, not only for thymocyte expansion but also supports hematopoiesis, suffers niche dysfunction due to HIV-1 infection. Continued presence of the CD34$^+$ progenitor stem cells in the infected niche seem to suffer due to exacerbation resulting from persistent virus mediated niche disruption via infection of thymocytes and consequent interactions and signaling network of the hubs. In this microenvironment it is possible that several ARGs could produce different outcomes. This is evident from our previous observation that CCR4 and CCR5- tropic HIV-1 produce variable kinetics of inhibition of hematopoiesis [60].

Less understood is the role of different lymphoid organs in regards to the diminution of T cells in the gastrointestinal niche which is involved in HIV infection clinical outcome especially the genetic markers that contribute to AIDS progression. The intestinal mucosal immune system is an important target of HIV-1 infection and contributes to disease progression. In addition distinct gene expression profiles correlate with clinical outcome [63]. In this regard we wonder if the variable time of outcome of AIDS following HIV infection is related to several innate unknown mechanisms operating in the host. Independent of these is the fact that patients with HIV do not get diagnosed immediately after infection and the time of diagnosis could be variable taking sometimes several years. For example, we had access to the San Francisco database [46] and several individuals recruited for follow-up were first evaluated from 0 to 11 years and therefore the definition of long-term non-progression can only be analyzed after years of the date of known infection. In addition, some of such patients were censored after they had begun treatment. Only in those individuals followed from the time of infection it would be possible to demonstrate the role of genetic markers in viremia, and their participation in producing decreased CD4 counts and/or conversion to AIDS.

The transcription factors such as STAT5A are involved in stem cell self-renewal that precedes multilineage differentiation of CD34+ progenitor stem cells [64-69]. The proto-oncogene of myeloproliferative leukemia also known as thrombopoietin (Tpo) receptor proto-oncogene, c-mpl, is known to promote multi-lineage pluripotent stem cell differentiation of the CD34+ progenitor cells [70-73]. Both STAT5 and c-mpl are important target genes for control and enhancement of stem cell self-renewal and multi-lineage differentiation to reduce or prevent cytopenias induced during HIV infection [74]. We wonder if some individuals have highly regulated expression of STAT5A/B and c-mpl genes in progenitor cells and if in such individuals CD4 cells are generated in higher numbers than in those without such a regulated niche.

IV. Discussion

It is necessary to investigate the role of NK cells in the innate immunity against HIV infection. It is now clear that there are at least three kinds of immune mechanisms involved in the control of viral infections including that of HIV-1: A) intrinsic innate immunity mediated by a group of major defenses against infection by retroviruses, Fv1 and TRIM5 inhibitors, proteins that target incoming retroviral capsids and the APOBEC3 class of cytidine deaminases that hypermutate and destabilize retroviral genomes. These are probably involved in the prevention of HIV-1 into cells and constitute the first level of protection from infection by specific cells, such as dendritic cells, including plasmocytoid dendritic cells. Of course many proteins such as cytokines, IL12, chemokines, and chemokine receptors CCR5 and CXCR4 are important. B) Genetic markers of immune effectors that are important in the outcome of HIV-1 infection; Natural killer, NK, receptors of two kinds, Ig-like such as KIRs and lectin-like such as NKG2 are involved. These two kinds of genetic markers have been described to be involved in the outcome of the HIV-I infection towards long term non-progression and AIDS. In this regard, two reports have described the role of HLA-Bw4/Bw4 as a marker for LTNP and related to that a subgroup of Bw4, isoleucine in position 80 of Bw4 interacting with a KIR receptor, 3DS1 was associated with LTNP [25, 26]. In unpublished work we have re-analyzed the data published before involving the importance of Bw4 in LTNP, discovered that methionine, Met/Met, in the second position of the leader peptide of HLA-B alleles is a marker for progression to AIDS [47]. NK receptors may be involved in the control of viremia; CD94/NKG2A inhibitory receptors [75-77] and the NKG2D stimulatory receptors not only present in NK cells but also in human CD8 lymphocytes [39, 78]. Therefore, NK and CTL activating and inhibitory signals can be provided by NKG2 receptor interaction with MICA and by CD94/NKG2A interaction with HLA-E molecules with the appropriate assembly of peptides [44]. However, the role of NKG2D/MICA in the progression to AIDS was not found [17]. The CD94/NKG2A-HLA-E pathway has not been studied related to HIV-1 infection outcome. It is important to mention that HLA leader peptide sequences with Met at P2 induce significant levels of HLA-E expression compared to those with Thr at P2, the latter failed to confer protection from NK lysis and the HLA-E/peptide complexes exhibited high affinity for soluble CD94/NKG2 molecules [45, 47]. Also, although HLA-A and C alleles encode HLA-E binding peptides with Met at P2, HLA-C alleles are poorly expressed on the surface of cells [39] and *nef* of HIV-1 down-regulates the cell surface expression of HLA-B and A but not C or HLA-E [40]. Furthermore, HLA-A alleles are rarely used to restrict CTL epitopes [48, 79]. In sum, in the environment of HIV-1 infected cells the HLA-leader peptides have more contribution to generate HLA-E binding peptides. This is supported by the reduced HLA-E poor expression by cells transfected with HLA-B*51 or B*58 [80, 81]. Therefore, it is possible that NK receptors are involved together with the recognition of CD8 T cell responses against the human immunodeficiency virus (HIV-1). Since HIV-I infection modulates the expression of IFNs that the cellular environment, niche, there would render greater contribution of HLA-B leader sequences to generate HLA-E binding peptides to interact with the NKG2A receptor as well as interaction with the cognate immune system [48]. Based on this scenario, we believe that the presence of Met at P2 produces inhibition of killing favoring HIV progression whereas, Thr at P2 produces loss of inhibition of killing favoring activating receptors of NK cells that negatively influence HIV infection outcome.

Therefore it is important to emphasize that among the multiple genes that limit the progression to AIDS [4] HLA-E molecules loaded with either methionine of threonine in the second position of leader peptides should be studied not only in regard to their influence in infection outcome but also in the innate protection from HIV-1 infection. Furthermore, we will also use a SCID-hu model where a combination of TRIM-5a, CCR5, with threonine in the second position of the HLA-B leader peptide will protect transplants from infection. C) Cognate immunity. There is a large literature of this subject related to viremia progression and future investigations are needed in this respect to identify genes of protection from infection by HIV. These should include genetic markers involved in the role of CD4/CD8 in HIV-1 infection together with those genetic factors involved in mechanisms of innate immunity against HIV-1 viral infection or disease progression, such as those involving NK receptors interacting with their ligands. This mechanism also explains the slow progression to AIDS in a subgroup of patients that maintain normal CD4+ T cell counts but demonstrate viremia even after 15 years of infection [46].

In unpublished studies by Flores, et al [46] previous published results were reanalyzed [25] with 39 individuals of known immune status that included controllers of viremia with a short follow up (not longer than 2 years) The HLA-B alleles were grouped according the Bw4 or Bw6 public specificity encoding epitopes determined by residues 79-83 at the carboxyl-terminal end of the alpha 1 helix (Table 1). Some of the Bw6 alleles have threonine in the second position of their leader peptide, for example HLA-B*1801, B*4101, B*35 alleles, B*40 alleles, and B*15 alleles. These results were corroborated analyzing 98 HIV infected individuals with follow-up of viral loads and CD4+ T cell counts with progression or lack of progression to immunodeficiency. TT was associated with long-term progressors and with long-term non-progressors that maintain normal CD4 counts beyond 15 years from the date of infection.

Figure 1 summarizes possible variability of CD34 cells in the population, different age at time of thymic involution.

This hypothesis can be studied as a cross-sectional study using the SCID mouse using peripheral blood and tissues available to investigate the number of CD4, CD8 and NK cells present at the time of short-term progression or long-term non-progression as a first step to study the role of NKG2A/ligand interaction in inhibition from killing, or actual killing, of HIV infected target cells as well as using the SCID-hu mouse model.

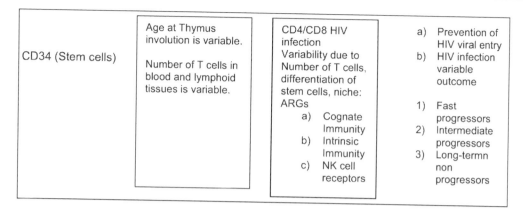

| CD34 (Stem cells) | Age at Thymus involution is variable.

Number of T cells in blood and lymphoid tissues is variable. | CD4/CD8 HIV infection
Variability due to Number of T cells, differentiation of stem cells, niche: ARGs
  a) Cognate Immunity
  b) Intrinsic Immunity
  c) NK cell receptors | a) Prevention of HIV viral entry
b) HIV infection variable outcome

1) Fast progressors
2) Intermediate progressors
3) Long-termn non progressors |

Figure 1. Diagram summarizes possible variability of CD34 cells in the population, different age at time of thymic involution.

V. Future Studies

Most studies of HIV infection outcome reported have described the importance of class I gene products particularly HLA-B. In this regard in a large number of HIV-1 infected individuals from southern Africa, it was reported that HLA-B alleles influence the potential co-evolution of HIV and HLA, providing advantage to pathogen defense mediated by CD8$^+$ T cells. These results were consistent with the findings in B-clade infected Caucasians with non-progression/low viral load or progression/high viral loads [48]. The authors claimed that the dominant effect for HLA-B alleles in HIV infection was unlikely due to differences in NK cell activation through the HLA-Bw4 motif. They mentioned the fact that it was reported that there is epistatic interaction between KIR3DS1 and some HLA-Bw4 alleles that mediated protection were entirely independent of this interaction [26]. Since there is a protective role of certain Bw4 alleles, B*2705, B*13 and B*44, as well as those that carry Ile-80, such a protective role should involve the mechanism described of NKG2A with the leader peptide threonine. See diagram. In both reports [26, 48], the authors did not consider the fact that there are two kinds of NK receptors, immunoglobulin receptors such as KIR3DS1 and the lectin receptors such as NKG2A. We have described herein that the alleles of Bw4, B*27 and B*57 as well as some Bw6 alleles carry threonine in the second position of their leader peptide that would interact with NKG2A. This mechanism should be studied together with the dominant role of HLA-B alleles influencing HIV specific CD8 T cell responses in the outcome of HIV infection. The use of cross-sectional studies has limitations of statistical power that are corrected when using longitudinal studies and this explains to some extent the major inconsistencies in published reports related to the HIV infection outcome. Perhaps the SCID mouse model could resolve some of these problems but this model also has limitations related to the fact that the experiments are limited to the particular tissues grafted and do not necessarily include the large number of ARGs present in individuals of the population at large [82, 83]. However, the engraftment of CD34$^+$ or even of CD133$^+$ cells give rise to multiple lineages of cells following spontaneous differentiation in vivo that may identify the source of ARGs.

Acknowledgement

Supported by NIH grants HL29583 and HL59838, and also from the Dana Farber Cancer Institute. P.K. is a recipient of NIH grant HL079846.

References

[1]   Fauci, A.S. HIV and AIDS: 20 years of science. *Nat. Med,* 2003, 9, 839−843.

[2]   O'Brien, S.J. and Moore, J. The effect of genetic variation in chemokines and their receptors on HIV transmission and progression to AIDS. *Immunol. Rev,* 2000, 177, 99−111.

[3]   Carrington, M. and O'Brien, S.J. The influence of HLA genotype on AIDS. *Ann. Rev. Med*, 2003 54, 535−551.

[4]   O'Brien SJ, Nelson GW. Human genes that limit *AIDS,* 2004, 36(6):565-74.

[5]   Winkler C, An P, O'Brien SJ Patterns of ethnic diversity among the genes that influence AIDS. *Hum. Mol. Gene.*, 2004,13 Spec No 1:R9-19.

[6]   Cao Y, Qin L, Zhang L, Safrit J, Ho DD.Virologic and immunologic characterization of long-term survivors of human immunodeficiency virus type 1 infection. *N. Engl. J. Med.*, 1995, 332(4):201-8.

[7]   Pantaleo G, Menzo S, Vaccarezza M, Graziosi C, Cohen OJ, Demarest JF, Montefiori D, Orenstein JM, Fox C, Schrager LK, et al.Studies in subjects with long-term nonprogressive human immunodeficiency virus infection. *N. Engl. J. Med.*, 1995, 332(4):209-16.

[8]   HRinaldo C, Huang XL, Fan ZF, Ding M, Beltz L, Logar A, Panicali D, Mazzara G, Liebmann J, Cottrill M, et al.igh levels of anti-human immunodeficiency virus type 1 (HIV-1) memory cytotoxic T-lymphocyte activity and low viral load are associated with lack of disease in HIV-1-infected long-term nonprogressors. *J. Virol.*, 1995, 69(9):5838-42.

[9]   Harrer T, Harrer E, Kalams SA, Barbosa P, Trocha A, Johnson RP, Elbeik T, Feinberg MB, Buchbinder SP, Walker BD.Cytotoxic T lymphocytes in asymptomatic long-term nonprogressing HIV-1 infection. Breadth and specificity of the response and relation to in vivo viral quasispecies in a person with prolonged infection and low viral load. *J. Immunol.*, 1996, 156(7):2616-23.

[10]   Kalams SA, Buchbinder SP, Rosenberg ES, Billingsley JM, Colbert DS, Jones NG, Shea AK, Trocha AK, Walker BD. Association between virus-specific cytotoxic T-lymphocyte and helper responses in human immunodeficiency virus type 1 infection, *J. Virol.*, 1999, 73(8):6715-20.

[11]   Lisziewicz J, Rosenberg E, Lieberman J, Jessen H, Lopalco L, Siliciano R, Walker B, Lori F.Control of HIV despite the discontinuation of antiretroviral therapy. *N. Engl. J. Med.*, 1999, 340(21):1683-4.

[12]   De Maria A, Ferraris A, Guastella M, Pilia S, Cantoni C, Polero L, Mingari MC, Bassetti D, Fauci AS, Moretta L. Expression of HLA class I-specific inhibitory natural killer cell receptors in HIV-specific cytolytic T lymphocytes: impairment of specific cytolytic functions, *Proc. Natl. Acad. Sci. USA*, 1997, 94(19):10285-8.

[13]   C. Servet, L. Zitvogel and A. Hosmalin Dendritic Cells in Innate Immune Responses Against HIV Pp.739-756. *Current Molecular Medicine*, 2 (8), 2002.

[14]   G. Herbein, A. Coaquette, D. Perez-Bercoff and G. Pancino Hosmalin. Macrophage Activation and HIV Infection: Can the Trojan Horse Turn into a Fortress? *Current Molecular Medicine*, 2002, 2 (8):723-738.

[15]   Rappocciolo G, Birch J, Ellis SA.Down-regulation of MHC class I expression by equine herpesvirus-1. *J Gen Virol*, 2003, 84(2), 293-300.

[16]   E.A. Trachtenberg and H.A. Erlich, A review of the role of the human leukocyte antigen (HLA) system as a host immunogenic factor influencing HIV transmission and progression to AIDS. In: B.T.K. Korber, C. Brander, B.F. Haynes, J.P. Moore, R.A. Koup, C. Kuiken, B.D. Walker and D.I. Watkins, Editors, HIV Molecular Immunology 2001, Theoretical Biology and Biophysics Group (2001), pp. 143–160.

[17]   Delgado JC, Leung JY, Baena A, Clavijo OP, Vittinghoff E, Buchbinder S, Wolinsky S, Addo M, Walker BD, Yunis EJ, Goldfeld AE. The -1030/-862-linked TNF promoter single-nucleotide polymorphisms are associated with the inability to control HIV-1 viremia. *Immunogenetics*, 2003, 55(7):497-501.

[18]   Folks TM, Clouse KA, Justement J, Rabson A, Duh E, Kehrl JH, Fauci AS.Tumor necrosis factor alpha induces expression of human immunodeficiency virus in a chronically infected T-cell clone. *Proc. Natl. Acad. Sci. USA*, 1989, 86(7):2365-8.

[19]   Marshall WL, Brinkman BM, Ambrose CM, Pesavento PA, Uglialoro AM, Teng E, Finberg RW, Browning JL, Goldfeld AE. Signaling through the lymphotoxin-beta receptor stimulates HIV-1 replication alone and in cooperation with soluble or membrane-bound TNF-alpha. *J. Immunol.*, 1999, 162(10):6016-23.

[20]   Poli G, Kinter A, Justement JS, Kehrl JH, Bressler P, Stanley S, Fauci AS.Tumor necrosis factor alpha functions in an autocrine manner in the induction of human immunodeficiency virus expression. *Proc. Natl. Acad. Sci. USA*, 1990, 87(2):782-5.

[21]   Gao, X. et al. Effect of a single amino acid change in MHC class I molecules on the rate of progression to AIDS. *N. Engl. J. Med*, 2001, 344, 1668−1675.

[22]   O'Brien, S.J., Gao, X. and Carrington, M. HLA and AIDS: A cautionary tale. *Trends Mol. Med.*, 2002, 7, 379−381.

[23]   Carrington, M. et al. HLA and HIV-1: Heterozygote advantage and B[*]35-Cw[*]04 disadvantage. *Science*, 1999, 283, 1748−1752.

[24]   Tang, J.M. et al. HLA class I homozygosity accelerates disease progression in human immunodeficiency virus type I infection. *AIDS Res. Hum.*, 1999, Retroviruses 15, 317−324.

[25]   Flores-Villanueva, P.O. et al. Control of HIV-1 viremia and protection from AIDS are associated with HLA-Bw4 homozygosity. *Proc. Natl. Acad. Sci. USA* 98, 2001, 5140−5145.

[26]   Martin MP, Gao X, Lee JH, Nelson GW, Detels R, Goedert JJ, Buchbinder S, Hoots K, Vlahov D, Trowsdale J, Wilson M, O'Brien SJ, Carrington M.Epistatic interaction between KIR3DS1 and HLA-B delays the progression to AIDS. *Nat. Genet*, 2002, 4:429-34.

[27]   Qing M, Li T, Han Y, Qiu Z, Jiao Y.Accelerating effect of human leukocyte antigen-Bw6 homozygosity on disease progression in Chinese HIV-1-infected patients. *J. Acquir. Immune Defic. Syndr*, 2006, 41(2):137-9.

[28]   Kaslow RA, Tand J, Dorak MT, Tang s, Musonda R, Karita E, Wilson C, allen S. Homozigosity for HLA-Bw4 is not associated with protection of HIV-1 infected persons in African ancestry. *Conference retroviruses opportunistic infect*, 2002, Feb 24-28;9: abstract no. 320-W.

[29]   Best, S., Le Tissier, P., Towers, G., and Stoye, J. P. (1996). Positional cloning of the mouse retrovirus restriction gene Fv1. Nature 382, 826-829.

[30]   Sheehy, A. M., Gaddis, N. C., Choi, J. D., and Malim, M. H. (2002). Isolation of a human gene that inhibits HIV-1 infection and is suppressed by the viral Vif protein. Nature 418, 646-650.

[31]   Stremlau, M., Owens, C. M., Perron, M. J., Kiessling, M., Autissier, P., and Sodroski, J. (2004). The cytoplasmic body component TRIM5alpha restricts HIV-1 infection in Old World monkeys. Nature 427, 848-853.

[32]   Li, Y., Li, X., Stremlau, M., Lee, M., and Sodroski, J. (2006). Removal of arginine 332 allows human TRIM5alpha to bind human immunodeficiency virus capsids and to restrict infection. J Virol 80, 6738-6744.

[33]   Yap, M. W., Nisole, S., and Stoye, J. P. (2005). A single amino acid change in the SPRY domain of human Trim5alpha leads to HIV-1 restriction. Curr Biol 15, 73-78.

[34]   Javanbakht H, An P, Gold B, Petersen DC, O'Huigin C, Nelson GW, O'Brien SJ, Kirk GD, Detels R, Buchbinder S, Donfield S, Shulenin S, Song B, Perron MJ, Stremlau M, Sodroski J, Dean M, Winkler C. Effects of human TRIM5alpha polymorphisms on antiretroviral function and susceptibility to human immunodeficiency virus infection, *Virology*, 2006, 354(1):15-27.

[35]   Winkler, C.A. and O'Brien, S.J. *AIDS restriction genes in human ethnic groups: An assessment. in AIDS in Africa* (eds. Essex, M., Mboup, S., Kanki, P.J., Marlink, R. and Tlou, S.D.) 2nd edn. Kluwer Academic, New York, 2002.

[36]   Speelmon EC, Livingston-Rosanoff D, Li SS, Vu Q, Bui J, Geraghty DE, Zhao LP, McElrath MJ.Genetic association of the antiviral restriction factor TRIM5alpha with human immunodeficiency virus type 1 infection. *J. Virol.*, 2006, 80(5):2463-71.

[37]   Scott-Algara D, Truong LX, Versmisse P, David A, Luong TT, Nguyen NV, Theodorou I, Barre-Sinoussi F, Pancino G.Cutting edge: increased NK cell activity in HIV-1-exposed but uninfected Vietnamese intravascular drug users. *J. Immunol.*, 2003, 171(11):5663-7.

[38]   Collins KL, Chen BK, Kalams SA, Walker BD, Baltimore D.HIV-1 Nef protein protects infected primary cells against killing by cytotoxic T lymphocytes. *Nature*, 1998, 391(6665):397-401.

[39]   McCutcheon JA, Gumperz J, Smith KD, Lutz CT, Parham P.Low HLA-C expression at cell surfaces correlates with increased turnover of heavy chain mRNA. *J. Exp. Med.*, 1995,181(6):2085-95.

[40]   Cohen GB, Gandhi RT, Davis DM, Mandelboim O, Chen BK, Strominger JL, Baltimore D. The selective downregulation of class I major histocompatibility complex proteins by HIV-1 protects HIV-infected cells from NK cells. *Immunity*, 1999,10(6):661-71.

[41]   Liu K, Kao KJ. Mechanisms for genetically predetermined differential quantitative expression of HLA-A and -B antigens. *Hum. Immunol.* 2000 Aug; 61(8):799-807.

[42]   Katsuyama Y, Ota M, Ando H, Saito S, Mizuki N, Kera J, Bahram S, Nose Y, Inoko H.Sequencing based typing for genetic polymorphisms in exons, 2, 3 and 4 of the MICA gene. *Tissue Antigens*, 1999, 54(2):178-84.

[43]   Natarajan K, Dimasi N, Wang J, Mariuzza RA, Margulies DH.Structure and function of natural killer cell receptors: multiple molecular solutions to self, nonself discrimination. *Annu. Rev. Immunol.*, 2002, 20:853-85.

[44]   Phillips JH, Gumperz JE, Parham P, Lanier LL. Superantigen-dependent, cell-mediated cytotoxicity inhibited by MHC class I receptors on T lymphocytes. *Science*, 1995, 268(5209):403-5.

[45]   Borrego F, Ulbrecht M, Weiss EH, Coligan JE, Brooks AG.Recognition of human histocompatibility leukocyte antigen (HLA)-E complexed with HLA class I signal sequence-derived peptides by CD94/NKG2 confers protection from natural killer cell-mediated lysis. *J. Exp. Med.*, 1998, 187(5):813-8.

[46]   Flores-Villanueva, Yunis E, Buchbinder S, Vittinghoff E, walker B. Association of two copies of HLA-B alleles encoding HLA-E binding peptides with Threonine at postion 2 with control of viremia and progression to AIDS. Unpublished.

[47]   Brooks AG, Borrego F, Posch PE, Patamawenu A, Scorzelli CJ, Ulbrecht M, Weiss EH, Coligan JE.Specific recognition of HLA-E, but not classical, HLA class I molecules by soluble CD94/NKG2A and NK cells. *J. Immunol.*, 1999, 162(1):305-13.

[48]   Kiepiela P, Leslie AJ, Honeyborne I, Ramduth D, Thobakgale C, Chetty S, Rathnavalu P, Moore C, Pfafferott KJ, Hilton L, Zimbwa P, Moore S, Allen T, Brander C, Addo MM, Altfeld M, James I, Mallal S, Bunce M, Barber LD, Szinger J, Day C, Klenerman P, Mullins J, Korber B, Coovadia HM, Walker BD, Goulder PJ. Dominant influence of HLA-B in mediating the potential co-evolution of HIV and HLA. *Nature*, 2004, 432(7018):769-75.

[49]   Lanier LL.Natural killer cells: from no receptors to too many. *Immunity*, 1997, (4):371-8.

[50]   Miller JD, Weber DA, Ibegbu C, Pohl J, Altman JD, Jensen PE. Analysis of HLA-E peptide-binding specificity and contact residues in bound peptide required for recognition by CD94/NKG2. *J. Immunol.*, 2003, 171(3):1369-75.

[51]   WWW.anthonynolan.com/HIGseq/pep).

[52]   Koka, PS; Reddy, ST. Cytopenias in HIV infection: Mechanisms and alleviation of hematopoietic inhibition. *Curr. HIV Res.*, 2004, 2, 275-282.

[53]   Miles, SA; Mitsuyasu, RT; Moreno, J; Baldwin, G; Alton, NK; Souza, L; Glaspy, JA. Combined therapy with recombinant granulocyte colony-stimulating factor and erythropoietin decreases hematologic toxicity from zidovudine. *Blood*, 1991, 77, 2109-2117.

[54]   Miles, SA; Lee, S; Hutlin, L; Zsebo, KM; Mitsuyasu, RT. Potential use of Human stem cell factor as adjunctive therapy for Human immunodeficiency virus-related cytopenias. *Blood*, 1991, 78, 3200-3208.

[55]   Ratner, L. Human immunodeficiency virus-associated autoimmune thrombocytopenic purpura: A review. *Am. J. Med.*, 1989, 86, 194-198.

[56]   Fauci, AS. Host factors and the pathogenesis of HIV-induced disease. *Nature*, 1996, 384, 529-534.

[57]   Harbol, AW; Liesveld, JL; Simpson-Haidaris, PJ; Abboud, CN. Mechanisms of cytopenia in human immunodeficiency virus infection. *Blood Rev.*, 1994, 8, 241-251.

[58]   Shen, H; Cheng, T; Preffer, FI; Dombkowski, D; Tomasson, MH; Golan, DE; Yang, O; Hofmann, W; Sodroski, JG; Luster, AD; Scadden, DT. Intrinsic Human immunodeficiency virus type 1 resistance of hematopoietic stem cells despite coreceptor expression. *J. Virol.*, 1999, 73, 728-737.

[59]   Koka, PS; Jamieson, BD; Brooks, DG; Zack, JA. Human immunodeficiency virus type-1 induced hematopoietic inhibition is independent of productive infection of progenitor cells in vivo. *J. Virol.* 1999, 73, 9089-9097.

[60]   Koka, PS; Fraser, JK; Bryson, Y; Bristol, GC; Aldrovandi, GM; Daar, ES; Zack, JA. Human immunodeficiency virus type 1 inhibits multilineage hematopoiesis in vivo. *J. Virol.*, 1998, 72, 5121-5127.

[61]   Jenkins, M; Hanley, MB; Moreno, MB; Wieder, E; McCune, JM. Human immunodeficiency virus-1 infection interrupts thymopoiesis and multilineage hematopoiesis in vivo. *Blood*, 1998, 91, 2672-2678.

[62]   Koka, PS; Kitchen, CM; Reddy, ST. Targeting c-Mpl for revival of human immunodeficiency virus type 1-induced hematopoietic inhibition when CD34+ progenitor cells are re-engrafted into a fresh stromal microenvironment in vivo. *J. Virol.*, 2004, 78, 11385-11392.

[63]   Sankaran S, Guadalupe M, Reay E, George MD, Flamm J, Prindiville T, Dandekar S. Gut mucosal T cell responses and gene expression correlate with protection against disease in long-term HIV-1-infected nonprogressors. *Proc. Natl. Acad. Sci. USA..* 2005, 102(28):9860-5.

[64]   Stier S, Cheng T, Dombkowski D, Carlesso N, Scadden DT. 2002. Notch1 activation increases hematopoietic stem cell self-renewal in vivo and favors lymphoid over myeloid lineage outcome. *Blood* 99: 2369-2378.

[65]   Pestina TI, Jackson CW. 2003. Differential role of Stat5 isoforms in effecting hematopoietic recovery induced by Mpl-ligand in lethally myelosuppressed mice. *Exp. Hematol.* 31: 1198-1205.

[66]   Schulze H, Ballmaier M, Welte K, Germeshausen M. 2000. Thrombopoietin induces the generation of distinct Stat1, Stat3, Stat5a and Stat5b homo- and heterodimeric complexes with different kinetics in human platelets. *Exp. Hematol.* 28: 294-304.

[67]   Zeng H, Masuko M, Jin L, Neff T, Otto KG, Blau CA. 2001. Receptor specificity in the self-renewal and differentiation of primary multipotential hematopoietic cells. *Blood* 98: 328-334.

[68]   Bradley HL, Couldrey C, Bunting KD. 2004. Hematopoietic-repopulating defects from STAT5-deficient bone marrow are not fully accounted for by loss of thrombopoietin responsiveness. *Blood* 103: 2965-2972.

[69]   Goncalves F, Lacout C, Villeval JL, Wendling F, Vainchenker W, Dumenil D. 1997. Thrombopoetin does not induce lineage-restricted commitment of Mpl-R expressing pluripotent progenitors but permits their complete erythroid and megakaryocytic differentiation. *Blood* 89:3544-3553.

[70]   Kaushansky K. 1998. Thrombopoietin and the hematopoietic stem cell. *Blood* 92:1-3.

[71]   Kaushansky K, Lin N, Grossman A, Humes J, Sprugel KH, Broudy VC. 1996. Thrombopoietin expands erythroid, granulocyte-macrophage, and megakaryocytic progenitor cells in normal and myelosuppressed mice. *Exp. Hematol.* 24:265-269.

[72]   Silvestris F, Cafforio P, Tucci M, Dammacco F. 2002. Negative regulation of erythroblast maturation by Fas-L+/TRAIL+ highly malignant plasma cells: a major pathogenic mechanism of anemia in multiple myeloma. *Blood* 99:1305-1313.

[73]   Solar GP, Kerr WG, Zeigler FC, Hess D, Donahue C, de Sauvage FJ, Eaton DL. 1998. Role of c-mpl in early hematopoiesis. *Blood* 92:4-10.

[74]   Schuringa JJ, Chung KY, Morrone G, Moore MAS. 2004. Constitutive activation of STAT5A promotes human hematopoietic stem cell self-renewal and erythroid differentiation. *J. Exp. Med.* 200: 623-635.

[75]   Mingari MC, Moretta A, Moretta L. Regulation of KIR expression in human T cells: a safety mechanism that may impair protective T-cell responses. *Immunol. Today.* 1998, 19(4):153-7.

[76]   Ponte M, Bertone S, Vitale C, Tradori-Cappai A, Bellomo R, Castriconi R, Moretta L, Mingari MC. Cytokine-induced expression of killer inhibitory receptors in human T lymphocytes. *Eur. Cytokine Netw.* 1998,9(3 Suppl):69-72.

[77]   Mingari MC, Ponte M, Bertone S, Schiavetti F, Vitale C, Bellomo R, Moretta A, Moretta L. HLA class I-specific inhibitory receptors in human T lymphocytes: interleukin 15-induced expression of CD94/NKG2A in superantigen- or alloantigen-activated CD8+ T cells. *Proc. Natl. Acad. Sci. USA.* 1998 Feb 3;95(3):1172-7.

[78]   Fodil N, Pellet P, Laloux L, Hauptmann G, Theodorou I, Bahram S. MICA haplotypic diversity. *Immunogenetics.* 1999, 49(6):557-60.

[79]   Lee N, Llano M, Carretero M, Ishitani A, Navarro F, Lopez-Botet M, Geraghty DE. HLA-E is a major ligand for the natural killer inhibitory receptor CD94/NKG2A. *Proc. Natl. Acad. Sci. USA.* 1998, 95(9):5199-204.

[80]   Llano M, Lee N, Navarro F, Garcia P, Albar JP, Geraghty DE, Lopez-Botet M. HLA-E-bound peptides influence recognition by inhibitory and triggering CD94/NKG2 receptors: preferential response to an HLA-G-derived nonamer. *Eur. J. Immunol.* 1998, 28(9):2854-63.

[81]   Andre P, Brunet C, Guia S, Gallais H, Sampol J, Vivier E, Dignat-George F. Differential regulation of killer cell Ig-like receptors and CD94 lectin-like dimers on NK and T lymphocytes from HIV-1-infected individuals. *Eur. J. Immunol.* 1999, 29(4):1076-85.

[82]   Sundell IB, Koka PS. Chimeric SCID-hu model as a human hematopoietic stem cell host that recapitulates the effects of HIV-1 on bone marrow progenitors in infected patients. *J. Stem Cells* 2006; 1(4): 283-300.

[83]   Melkus MW, Estes JD, Padgett-Thomas A, Gatlin J, Denton PW, Othieno FA, Wege AK, Haase AT, Garcia JV. Humanized mice mount specific adaptive and innate immune responses to EBV and TSST-1. *Nature Med.* 2006 12(11): 1316-1322.

In: Natural Killer T-Cells: Roles, Interactions and Interventions   ISBN: 978-1-60456-287-3
Editor: Nathan V. Fournier, pp. 173-187   © 2008 Nova Science Publishers, Inc.

Chapter 8

Optimizing the Use of Beta Glycolipids as NKT Cell Ligands

Tomer Adar, Ami Ben Ya'acov, Gadi Lalazar and Yaron Ilan[*]
Liver Unit, Hebrew University-Hadassah Medical Center, Jerusalem, Israel

Abstract

Natural killer T (NKT) cells are a subset of regulatory lymphocytes that have been implicated in the regulation of autoimmune processes. The major histocompatibility complex class I-like CD1d glycoprotein is a member of the CD1 family of antigen presenting molecules, and is responsible for the selection of NKT cells. CD1d presents a number of ligands to NKT or other CD1d-restricted T cells, including glycolipids from a marine sponge, bacterial glycolipids, normal endogenous glycolipids, tumor-derived phospholipids and glycolipids, and non-lipid molecules. Some of these glycolipid/phospholipid ligand-CD1d complexes have been crystallized, revealing their tertiary structures. Most available data is on alpha configuration ligands. Recently, β-glycolpids have emerged as a family of possible ligands for this subset of regulatory lymphocytes. The presentation of many of these molecules can have immune-potentiating effects, acting as adjuvant against infections or promoting more rapid clearance of certain viruses. β-glycolpids can also be protective against autoimmune diseases or cancer; they can also be deleterious. In this review we will discuss the potential use of these ligands against immune-mediated disorders.

Key Words: Ligands, glycolipids, NKT cells.

Abbreviations

NKT,	natural killer T cell;
β-GC,	β-glucosylceramide;
β-GalCer,	β-galactosylceramide;
α–GalCer,	alpha-galactosylceramide.

[*] E-mail address: ilan@hadassah.org.il, Fax: 972-2-6431021; Tel: 972-2-6778231

A. NKT Lymphocytes: Development and Classification

Most lymphocytes in the peripheral blood are recirculating immunocompetent cells that have developed the ability to recognize foreign antigens and respond accordingly. Natural killers cells (NK) lack surface molecules that are characteristic of B or T lymphocytes, and do not to require antigens for their activation[1]. Natural killer T (NKT) cells are a unique lineage of T cells that share properties with both NK cells and memory T cells[2]. The identification of NKT cells is based on further classification of T lymphocytes according to the cell surface markers they present, in which they resemble NK cells and T cells, and according to their biological function in terms of cytokine production and ligand reactivity[3]. This subset of lymphocytes may be either $CD4^+$ or double negative (CD4 and CD8 negative), express NK cell and memory T-cell markers, and is CD1d reactive. They are unique in their invariant Vα14-Jα18 TCR α-chain, and their T cell receptor (TCR) β-chain is biased towards Vβ8.2, Vβ2, and Vβ7. NKT cells are also unique in their glycolipid antigen reactivity and in their marked cytokine production[4].

About two decades ago, several studies in mice helped to identify a subset of αβ-T cell receptor $(αβ-TCR)^+$ lymphocytes, with the following properties: 1) double negative for CD8 and CD4; 2) intermediated level of TCR expression; and, 3) Vβ8 expression that was two to three-fold higher than in normal T cells[3]. Furthermore, additional reports demonstrated a subset of $αβ-TCR^+$ cells that expressed the NK1.1 marker (Nkrp1c or CD161c), which was considered to be expressed only by NK cells. Among the $NK1.1^+$ T cells, two further subgroups were identified: CD^{4+} and DN[3]. Additional characteristics were derived from analysis of the TCRs, in which both α and β chains followed certain patterns. The TCR-α chain is invariant and contains Vα14-Jα18, and the β-chain is biased towards Vβ8.2, Vβ2 and Vβ7[4]. Other characteristics of the $NK1.1^+αβ-TCR$ cells were derived from their biological interactions. Their development required β2-microglobulin despite a lack of CD8 expression, and was independent of major histocompatibility complex (MHC) class II expression[5]. NKT cells are reactive to the MHC class I-like molecule CD1d[6]. NKT cells were also found to be highly reactive to an α-structured-glycolipid known as α-galactosylceramide (α-GalCer)[7].

The double negative $αβ-TCR^+$ lymphocytes have been reported to produce high amounts of cytokines such as interleukin-4 (IL-4), interferon-γ (IFN-γ), and tumor necrosis factor (TNF)[8]. Furthermore, the fact that certain subsets of $CD4^+$ T cells in the thymus were found to produce large amounts of cytokines[9] while $NK1.1^-CD4^+$ thymocytes do not[10], suggests that the fraction of $CD4^+$ cells that are highly active in cytokine production are NKT cells. NKT cells were identified in mice, and similar cells were found in rats, primates[11], and humans[12,13]. In humans, the cells are characterized by the expression of invariant Vα24-Jα18 TCR α-chain and Vβ11 TCR β-chain[3].

One of the main challenges in the initial classification of NKT cells was due to the fact that the molecular characteristics used to identify NKT cells vary among mouse strains, and between mice and humans. For example, identifying NKT cells based on the expression of NK1.1 is appropriate when dealing with mouse strains, such as C57BL/6, that express the NK1.1 marker. However, NK1.1-positivity is irrelevant in other widely used strains, such as BLAB/c, that do not express NK1.1 [14]. Furthermore, in humans, the expression of the

NK1.1 homologue, CD161, is not limited to T cells with NKT cell characteristics[3]. Because these cells could not be named solely by their molecular characteristics, this subgroup of lymphocytes was variously called NK T or NKT cells, NK1.1+(like) T cells, natural T cells, iNKT cells, and Vα14 invariant (Vα14i) T cells[15]. Possibly the most accurate definition is, CD1d-dependent natural killer-like T cells; however, the term NKT cells is used most widely [16, 17].

NKT cells consist of three subgroups: 1) Vα14 invariant or Vα24 invariant NKT cells that express semi-invariant TCR encoded by Vα14 and Jα18 gene segments in mice, and Vα24 and Jα18 in humans that recognize CD1d[18]; 2) NKT cells that are also restricted to CD1d but use more diverse TCR than the first group[18]; and, 3) non-CD1d-restricted NKT cells that are not dependent on CD1d molecules[4].

NKT cell development begins in the thymus and continues in the periphery[4]. Common thymocytic precursors, positive for both CD4 and CD8 (double positive), undergo random TCR gene rearrangement resulting in the expression Vα14-Jα18 in conjunction with either Vβ8.2, Vβ7 or Vβ2, leading to CD1d-dependent selection and the formation of the NKT lineage[4]. Expression of NK1.1 generally follows migration out of the thymus, but can precede it[3]. Unique molecular interactions govern NKT development and emigration from the thymus, yet it is unclear whether there is a single autologous antigen responsible for both positive selection and peripheral activation of these regulatory lymphocytes[19]. Mature NKT cells make up majority of peripheral NKT cells, however, a steady and sizable number of immature NKT cells migrate from the thymus to the periphery. These immature cells are functional but are likely to behave differently from their mature counterparts[3]. An example of the importance of TCR or CD1d in the development of NKT cells is demonstrated by the fact that disruption of TCR-α chain or CD1d results in the selective deficiency of NKT cells, though other types of lymphocytes remain intact[20].

In mice, NKT cells are most commonly found in the liver and account for 30% to 50% of hepatic lymphocytes[21], However, NKT cells can also be found wherever regular T cells are found, and account for 20% to 30% of bone marrow lymphocytes, 10% to 20% thymic lymphocytes, and constitute smaller proportions in the lung (7%), blood (4%), spleen (3%), and lymph nodes (0.3%)[15]. In humans, the exact distribution of NKT cells in different organs has not been determined; however, in the liver the fraction of NKT cells varies between 4% and 20%[22, 23].

B. NKT Lymphocyte Function

NKT cells, CD1d restricted NKT cells in particular, have several functions, some characterized as direct effector functions and others as immunomodulatory functions[29,30]. Upon activation, NKT cells rapidly produce different cytokines including IFN-γ and IL-4, express activation markers (e.g. CD69), down regulate NK1.1 surface antigen, and disappear from tissues in which they are usually found, probably due to activation-induced cell death[29, 31-34]. Activated NKT cells display cytotoxic capabilities, mediated by Fas, granzyme A/B, perforin, and granulysin[35, 36]. This process eventually results in the activation of both the innate and the adaptive immune system[28, 37-39]. One study reported that co-administration of α-GalCer with an antigen shifts the antigen specific T cell response towards the T_h1 cytokine profile[40], whereas other reports have demonstrated a T_h2

polarization following α-GalCer[41] administration. The ability of NKT cells to generate both T_h1 and T_h2 responses indicates their importance as immunoregulatory cells and the complexity of their modulatory machinery: different responses can be generated by the same ligands[16, 42]. In addition, NKT activation leads to NK cell activation, proliferation of memory $CD4^+$ and $CD8^+$ T cells, or expression of the early activation marker CD69 on the surface of T and B cells[43, 44].

NKT lymphocytes have been implicated in the regulation of autoimmune processes in both mice and humans[45]. Reduced numbers and defective function of NKT lymphocytes were demonstrated in non-obese diabetic (NOD) mice. Transplantation of NKT lymphocytes, introduction of a Vα14Jα281 transgene onto the NOD background, or activation of NKT lymphocytes by administration of α-GalCer, ameliorates diabetes in this model[46, 47]. Administration of α-GalCer and OCH, a sphingosine-truncated analogue of α-GalCer, had protective effects in experimental autoimmune encephalomyelitis (EAE) and collagen-induced arthritis in mice, respectively[48]. In these models, disease amelioration was associated with a shift in the immune balance from a pathologic T_h1 type response towards a protective T_h2 type response. Altered number or function of NKT cells have also been described in several human autoimmune diseases[48]. Patients with systemic lupus erythematosus, scleroderma, diabetes, multiple sclerosis, Sjogren syndrome, and rheumatoid arthritis were found to have lower numbers of peripheral blood NKT cells as compared to healthy subjects. Invariant Vα24JαQ cells were found to have a regulatory role in human subjects with scleroderma[46].

C. Ligands for NKT Regulatory Cells

Several glycolipids and phospholipids derived from mammalian, bacterial, protozoan, and plant species have been identified as possible natural ligands for NKT cells[49]. Some have been crystallized in CD1d-bound forms, revealing their tertiary structures[18]. The two main subgroups of NKT cells recognize CD1d molecules. CD1 molecules, which present nonpeptide, mostly lipid antigens, resemble MHC molecules, especially MHC class I molecules, since both their heavy chains bind non-covalently with β2-microglobulin[50, 51]. In humans, CD1 molecules are encoded on chromosome 1, with four possible isoforms: CD1a, CD1b, CD1c, and CD1d. Sequence homology allows further classification into two groups. Group 1 consists of CD1a, CD1b, and CD1c, and does not exist in mice. Group 2 consists of CD1d, and exists both in mice and humans[51]. CD1d molecules are constitutively expressed by different cells, including antigen presenting cells such as macrophages and dendritic cells, and B and T lymphocytes in the thymus and liver[52, 53]. CD1d molecules present antigens, mainly glycolipids, to NKT cells[6]. Several potential natural glycolipids have been suggested to activate NKT cells, including glycolipids such as glycosylphosphatidylinositol [54], glycosphingolipids such as isoglobotirhexosylceramide[55] and α-glucuronsylceramide[56, 57], and phospholipids such as phosphatidylcholine and phosphatidylinositol[58, 59]. It is noteworthy that not all naturally occurring lipid ligands are stimulatory. Gangliotriaosylceramide, which is secreted by certain murine T cell lymphoma cells, inhibits CD1d mediated antigen presentation[60]. The semi-invariant αβ–TCRs can recognize iGb3, a mammalian glycosphingolipid, and a microbial α-glycuronylceramides found in the cell walls of Gram-negative, lipopolysaccharide-negative bacteria[61]. iGb3 was proposed as one of the candidates recognized by NKT cells under pathological conditions such as cancer and auto-immune disease[61,62]. The dual recognition of self and microbial ligands underlies the innate-like antimicrobial functions mediated by CD40L induction, and massive cytokine and chemokine release by NKT cells[63]. There are multiple ways in which NKT cells are activated during microbial infection, and some may be associated with proteins that control lipid metabolism[64]. During infectious assault, the presentation of a neo-self glycolipid by antigen presenting cells activates iNKT cells, which release pro-inflammatory or anti-inflammatory cytokines and jump-start the immune system[66].

In 1994, a Japanese pharmaceutical company extracted several glycosphingolipids from the Okinawan marine sponge *Agelas mauritanus*[67]. These agelasphins, which consisted of D-galactose and ceramide, showed anti-tumor activity in mice[68]. However, only small amounts of agelasphins are contained in each sponge, so the modified derivative α-GalCer was produced[69]. The only efficient method for selectively stimulating NKT cells *in vivo* is via the sea sponge-derived agent α–GalCer[67]. Multimers of CD1d1-α–GalCer and α–GalCer analogue-loaded complexes demonstrate cooperative engagement of the Vα14Jα18 iNKT cell receptor[68, 70]. Administration of α–GalCer causes potent activation of NKT cells, rapid and robust cytokine production, and activation of a variety of cells of the innate and adaptive immune systems[67]. Administration of α–GalCer induces the secretion of both IL-4 and IFN-γ. Repeated administration favors the production of T_h2 cytokines[48, 71]. OCH is a unique analogue that selectively stimulates NKT to produce IL-4 and IL-10, and may be more beneficial for suppression of T_h1-mediated diseases.

Despite its marine origin, α-GalCer is a potent ligand for NKT cells after binding to CD1d expressed by an antigen presenting cells. The CD1d/α-GalCer complex is recognized by NKT cells via Vα14i TCR in mice and Vα24i TCR in humans[72]. NKT cell activation results in the production of both T_h1 and T_h2 cytokines; this has been observed in both mice and humans[27, 30]. The necessity of the CD1d molecule and the specificity of Vα14i-NKT cells in this process have been demonstrated by the absence of the reaction in CD1d- and Vα14i NKT-deficient mice[27]. CD1d-mediated recognition of α-GalCer is highly conserved. Brossay et al. (1998) showed that human NKT cells recognize α-GalCer presented by mouse CD1d molecules and vice versa[73]. However, α-GalCer has been shown to be hepatotoxic in mice, limiting its use in human testing[34]. Our understanding of the processes involved stimulatory or inhibitory NKT pathways has been advanced through investigations using several analogs of α-GalCer in different models: OCH, Yamamura et al. (2001) studied the effects of an α-GalCer analog with a shortened sphingosine chain (named OCH), and found it to be even more effective in EAE models than α-GalCer, provoking increased IL-4 production[74]; KRN 7000, an analog with di-unsaturated 20 carbon chain (C20:2), was reported to stimulate IL-4 production yet inhibit IFN-γ[75]; α-C-GalCer, a C-glycoside analog, was reported to generate a longer lasting reaction, with a clearer T_{h1} response[76, 77]. In addition to α-GalCer and its analogs, β-GalCer (C12) has also been reported to be a CD1d ligand capable of stimulating NKT cells[78, 79].

D. Immune Modulatory Activities of α-GalCer and Its Analogues

Investigators have studied the role of activated NKT cells in various types of pathologic conditions, including infectious, autoimmune, and neoplastic[46, 80]. 1) Hepatitis B Viral Infection: In hepatitis B virus transgenic mice, NKT cell activation abolished HBV replication[81]. 2) Tuberculosis: CD1 restricted T cells recognize both pure lipid mycobacterium antigens presented by CD1 and those processed by macrophages infected by *Mycobacterium tuberculosis*[82]. Several activated CD1-restricted T cells with granulysin mediated bactericidal capabilities can lyse uninfected macrophages [83]. The role of NKT cells in granuloma formation has been demonstrated[84]; however, lipid antigen processing and presentation in mycobacterium lung infection was not efficient enough to induce NKT cell activation[82]. Alpha-GalCer-mediated activation was protective in this setting. 3) Cryptococcus infections: *Cryptococcus neoformans* is a ubiquitous fungal pathogen. Among the host factors determining susceptibility and vulnerability is the balance between T_h1 and T_h2 responses; T_h1 predominant responses are protective[85]. Alpha-GalCer-induced NKT cell activation has been shown to increase IFN-γ production and augment local host resistance to *C. neoformans* infection[86, 87]. 4) Malaria: Administration of α-GalCer enhances malaria immunity in mice who had received sub-optimal doses of irradiated sporozoites or recombinant viruses expressing viral antigens[77]. Interestingly, α-GalCer had a maximal adjuvant effect when co-administered with the antigens, and showed no adjuvant effect when given two days prior to antigen administration[77]. 5) NKT cells and encephalitis: In an EAE model , α-GalCer-induced NKT cell activation potentiated or prevented disease, depending on the NKT reaction[88]. IFN-γ secretion was associated with disease exacerbation, whereas IL-4 production was protective against EAE. Prior immunization with α-GalCer resulted in increased IL-4 secretion. 6) Malignancy: The role of NKT cells in the setting of neoplastic disease remains under investigation. Having demonstrated that NKT cells are capable of causing tumor regression via IL-13 mediated inhibition of tumor specific cytotoxic T lymphocytes[89], it was suggested that NKT cells might normally inhibit tumor immunity. In contrast, the absence of NKT cells was associated with an increased risk of tumorogenesis[25]. Crowe et al. (2002) asked why the potential for NKT cells to generate an anti-tumor reaction depended on exogenous stimuli such as IL-12 and α-GalCer for some tumors (B16F10 melanoma), but not others (sarcomas)[20]. They hypothesized that different tumors display different CD1d binding glycolipid antigens. 7) Diabetes: In NOD mice, in which diabetes is mediated by T_h1 cells, α-GalCer-activated NKT cells can prevent pancreatic islet β cell destruction[90, 91].

Several studies have suggested that disease-target antigens can serve as NKT cell ligands. In NOD mice, NKT-dependent amelioration of diabetes is noted following vaccination with disease-target antigens such as insulin, GAD65, and sterptozotocin-treated islets[92]. Adoptive transfer of *ex vivo* disease target antigen-exposed NKT cells alleviated immune-mediated colitis[93,94]. NKT lymphocytes exposed to tumor antigens suppressed tumor growth[95]. In asthma, NKT cells migrate from the thymus, spleen, liver, and bone marrow into blood vessels, and concentrate in airway bronchi mucosa[96]. This recruitment is dependent on CCR9 expression and engagement of CCL25/CCR9 in response to target antigens. Circulating asthmatic NKT cells express high levels of the T_h1 cytokines IFN-γ, and once they reached airway epithelium and are exposed to disease-target antigens, most NKT cells shift to T_h2-bias, and express high levels of IL-4, IL-13[96].

E. β-Glycolipids as NKT Ligands

Alpha-anomeric D-glycosylceramides have not been detected in mammals. Several recent studies have suggested that endogenous β-structured glycolipids may be potent NKT ligands[97]. Beta-structured glycolipids are normal constituents of cell membranes[98, 99]. β-glucosylceramide (β-GC) is a naturally occurring glycolipid that is a metabolic intermediate in glycosphingolipid anabolic and catabolic pathways. Its synthesis from ceramide is catalyzed by glucosylceramide synthase[97]. Circumstantial evidence pointing to β-GC involvement in NKT cell regulation can be derived from patients with Gaucher disease. In Gaucher disease, the most common lysosomal storage disease, decreased activity of glucosylceramide synthase results in elevated serum β-GC levels[100]. Interestingly, patients with Gaucher disease have altered humoral and cellular immune profiles, including altered NKT cells number and function. Furthermore, findings in patients with Gaucher disease suggest a direct effect of β-GalCer on cellular membranes. Some patients have increased red blood cell aggregation[101] due in part to changes in cellular membranes properties.

Following binding to CD1d, β-GC has two equally efficient pathways of action: a direct pathway, resulting from β-GC binding and presentation, and a second pathway, resulting from the inhibition of α-GalCer mediated activation[78]. Similar to α-GalCer activation, β-D-GalCer leads to a reduction in NKT cells. However, unlike α-GalCer, β-GalCer is a poor inducer of IFN-γ, TNF-α, GM-CSF, and IL-4 gene expression[78]. *In vitro*, CD1d-bound β-GC inhibits NKT cell activation by α-GalCer[78]. Glucosylceramide synthase deficiency leads to defective ligand presentation by CD1d, inhibiting NKT activation. β-GalCer-deficient mice exhibit normal NKT development and function, and cells from these animals can stimulate NKT hybridomas[102]. In striking contrast, the same hybridomas fail to react to CD1d1 expressed by a β-D-glucosylceramide (β-D-GlcCer)-deficient cell line. Human β-D-GlcCer synthase cDNA transfer restores the recognition of mutant cells expressing CD1d1 by Vα14Jα18 NKT hybridomas. Suppression of β-D-GlcCer synthesis inhibits antigen presentation to iNKT cells; however, β-D-GlcCer itself is unable to activate NKT hybridomas[102]. β-D-GalCer (C12) efficiently diminished the number of detectable NKT cells *in vivo* without inducing cytokine expression. Binding studies have demonstrated that both α–GalCer and β-D-GalCer were equally efficient in reducing the number of NKT cells. However, in contrast to α–GalCer, β-D-GalCer (C12) is a poor inducer of IFN-γ, TNF-α, GM-CSF, and IL-4 gene expression[78].

In addition to CD1d-binding mediated mechanisms, it was proposed that β-glycolipids affect NKT cell activation via alteration of cell membrane properties, specifically those of lipid rafts[103]. Lipid rafts are highly dynamic submicroscopic assemblies enriched in sphingolipids and cholesterol. Alteration in raft properties may impair raft receptor localization without necessarily inhibiting ligand-receptor binding[104]. Raft disruption has been shown to inhibit IL-6/STAT3 and IFN-γ/STAT1 signaling[104]. CD1d is also localized in lipid rafts[105], and disruption of the lipid rafts can inhibit NKT activation without impairing CD1d-ligand binding[105]. The administration of naturally occurring β-glycolipids can alter lipid raft composition and structure, thereby affecting intracellular signaling machinery[66, 78]. It has been reported that β-GalCer (C12) stimulates NKT cells[79]. β-glycolipids seem to have a different effect on NKT cells than does α-GalCer, inhibiting NKT cell proliferation without stimulating cytokine expression[106].

The effect of β-glycolipids in immune-mediated disorders has been studied in several animal models. 1) Immune mediated colitis: In a murine T_h1-mediated colitis model, β-GalCer generated a T_h2 response, associated with colitis alleviation[42]. 2) Immune mediated hepatitis: β-GalCer alleviated concanalvalin-A induced hepatitis in mice, associated with decreased serum IFN-γ levels and reduced expression of the transcription factor STAT1[106]. 3) Hepatocellular carcinoma: In mice with hepatocellular carcinoma, β-GalCer administration suppressed tumor growth and improved survival, associated with a T_h1 immune shift[42]. 4) Graft versus host disease: β-GalCer treatment was effective in alleviating semiallogeneic acute and chronic graft versus host disease in mice[107].

One of the most intriguing aspects of NKT cells is their plasticity. This subset of lymphocytes generates both T_h1 and T_h2 responses upon activation; however, the mechanisms and consequences of their plasticity remain under investigation. First, it is unclear whether or not this plasticity is ligand mediated. Different immune responses are associated with typical mediators; however, the same ligand can generate different immune type responses in different immune microenvironment settings. In light of differing effects of a given ligand *in vivo* and *in vitro*, the net effect of NKT activation may not result from the binding of a single ligand, but result from the sum of the effects of a variety of mediators. Furthermore, organ-specific factors may play a role in NKT plasticity, with different responses generated in different organs by an identical stimulus. Alternatively, identical original stimuli may be reach NKT cells via different pathways resulting from different antigen presenting cells[16, 107]. NKT cells are a heterogeneous population of cells that differ from one another in their CD1d reactivity and CD expression, contributing to plasticity. Apart from inherent heterogeneity between different NKT populations, changes in cellular membranes with altered lipid raft properties affect raft-bound receptors (such as CD1d), and may add to the variety of responses. NKT plasticity may be an expression of different immunologic reaction profiles, dictated by genetic heterogeneity. NKT plasticity may thus be a result of several of the factors mentioned above, with CD1d ligands being the last link in a chain of factors determining the final response profile.

The overall objective of ongoing research is to determine the NKT ligand-dependent signaling pathways in order to optimize the use of different ligands as immunomodulatory agents. Understanding the relationship between NKT intracellular signaling pathways and ligand structure, and the dependency on the CD1d system and on specific NKT raft micro domains, will contribute to the development of NKT-based immunotherapy.

References

[1]　　　Arina A, Murillo O, Dubrot J, Azpilikueta A, Alfaro C, Perez-Gracia JL, Bendandi M, Palencia B, Hervas-Stubbs S, Melero I. Cellular liaisons of natural killer lymphocytes in immunology and immunotherapy of cancer. *Expert Opin Biol Ther* 2007;7:599-615.

[2]   Bendelac A, Rivera MN, Park SH, Roark JH. Mouse CD1-specific NK1 T cells: development, specificity, and function. *Annu Rev Immunol* 1997;15:535-62.

[3]   Godfrey DI, MacDonald HR, Kronenberg M, Smyth MJ, Van Kaer L. NKT cells: what's in a name? *Nat Rev Immunol* 2004;4:231-7.

[4]   Bendelac A, Savage PB, Teyton L. The biology of NKT cells. *Annu Rev Immunol* 2007;25:297-336.

[5]   Ohteki T, MacDonald HR. Major histocompatibility complex class I related molecules control the development of CD4+8- and CD4-8- subsets of natural killer 1.1+ T cell receptor-alpha/beta+ cells in the liver of mice. *J Exp Med* 1994;180:699-704.

[6]   Bendelac A, Lantz O, Quimby ME, Yewdell JW, Bennink JR, Brutkiewicz RR. CD1 recognition by mouse NK1+ T lymphocytes. *Science* 1995;268:863-5.

[7]   Brigl M, Brenner MB. CD1: antigen presentation and T cell function. *Annu Rev Immunol* 2004;22:817-90.

[8]   Zlotnik A, Godfrey DI, Fischer M, Suda T. Cytokine production by mature and immature CD4-CD8- T cells. Alpha beta-T cell receptor+ CD4-CD8- T cells produce IL-4. *J Immunol* 1992;149:1211-5.

[9]   Hayakawa K, Lin BT, Hardy RR. Murine thymic CD4+ T cell subsets: a subset (Thy0) that secretes diverse cytokines and overexpresses the V beta 8 T cell receptor gene family. *J Exp Med* 1992;176:269-74.

[10]   Arase H, Arase N, Nakagawa K, Good RA, Onoe K. NK1.1+ CD4+ CD8- thymocytes with specific lymphokine secretion. *Eur J Immunol* 1993;23:307-10.

[11]   Motsinger A, Azimzadeh A, Stanic AK, Johnson RP, Van Kaer L, Joyce S, Unutmaz D. Identification and simian immunodeficiency virus infection of CD1d-restricted macaque natural killer T cells. *J Virol* 2003;77:8153-8.

[12]   Dellabona P, Padovan E, Casorati G, Brockhaus M, Lanzavecchia A. An invariant V alpha 24-J alpha Q/V beta 11 T cell receptor is expressed in all individuals by clonally expanded CD4-8- T cells. *J Exp Med* 1994;180:1171-6.

[13]   Porcelli S, Yockey CE, Brenner MB, Balk SP. Analysis of T cell antigen receptor (TCR) expression by human peripheral blood CD4-8- alpha/beta T cells demonstrates preferential use of several V beta genes and an invariant TCR alpha chain. *J Exp Med* 1993;178:1-16.

[14]   Godfrey DI, McConville MJ, Pellicci DG. Chewing the fat on natural killer T cell development. *J Exp Med* 2006;203:2229-32.

[15]   Godfrey DI, Hammond KJ, Poulton LD, Smyth MJ, Baxter AG. NKT cells: facts, functions and fallacies. *Immunol Today* 2000;21:573-83.

[16]   Godfrey DI, Kronenberg M. Going both ways: immune regulation via CD1d-dependent NKT cells. *J Clin Invest* 2004;114:1379-88.

[17]   Berzins SP, Smyth MJ, Godfrey DI. Working with NKT cells--pitfalls and practicalities. *Curr Opin Immunol* 2005;17:448-54.

[18]   Tsuji M. Glycolipids and phospholipids as natural CD1d-binding NKT cell ligands. *Cell Mol Life Sci* 2006;63:1889-98.

[19]   Kronenberg M, Engel I. On the road: progress in finding the unique pathway of invariant NKT cell differentiation. *Curr Opin Immunol* 2007;19:186-93.

[20]   Crowe NY, Smyth MJ, Godfrey DI. A critical role for natural killer T cells in immunosurveillance of methylcholanthrene-induced sarcomas. *J Exp Med* 2002;196 119-27.

[21]   Crispe IN. Hepatic T cells and liver tolerance. *Nat Rev Immunol* 2003;3:51-62.

[22]   Doherty DG, Norris S, Madrigal-Estebas L, McEntee G, Traynor O, Hegarty JE, O'Farrelly C. The human liver contains multiple populations of NK cells, T cells, and CD3+CD56+ natural T cells with distinct cytotoxic activities and Th1, Th2, and Th0 cytokine secretion patterns. *J Immunol* 1999;163:2314-21.

[23]   Ishihara S, Nieda M, Kitayama J, Osada T, Yabe T, Ishikawa Y, Nagawa H, Muto T, Juji T. CD8(+)NKR-P1A (+)T cells preferentially accumulate in human liver. *Eur J Immunol* 1999;29:2406-13.

[24]   Arase H, Arase N, Kobayashi Y, Nishimura Y, Yonehara S, Onoe K. Cytotoxicity of fresh NK1.1+ T cell receptor alpha/beta+ thymocytes against a CD4+8+ thymocyte population associated with intact Fas antigen expression on the target. *J Exp Med* 1994;180:423-32.

[25]   Smyth MJ, Thia KY, Street SE, Cretney E, Trapani JA, Taniguchi M, Kawano T, Pelikan SB, Crowe NY, Godfrey DI. Differential tumor surveillance by natural killer (NK) and NKT cells. *J Exp Med* 2000;191:661-8.

[26]   Hong S, Scherer DC, Singh N, Mendiratta SK, Serizawa I, Koezuka Y, Van Kaer L. Lipid antigen presentation in the immune system: lessons learned from CD1d knockout mice. *Immunol Rev* 1999;169:31-44.

[27]   Kawano T, Cui J, Koezuka Y, Toura I, Kaneko Y, Motoki K, Ueno H, Nakagawa R, Sato H, Kondo E, Koseki H, Taniguchi M. CD1d-restricted and TCR-mediated activation of valpha14 NKT cells by glycosylceramides. *Science* 1997;278:1626-9.

[28]   Van Kaer L. NKT cells: T lymphocytes with innate effector functions. *Curr Opin Immunol* 2007;19:354-64.

[29]   Van Kaer L. Natural killer T cells as targets for immunotherapy of autoimmune diseases. *Immunol Cell Biol* 2004;82:315-22.

[30]   Burdin N, Brossay L, Kronenberg M. Immunization with alpha-galactosylceramide polarizes CD1-reactive NK T cells towards Th2 cytokine synthesis. *Eur J Immunol* 1999;29:2014-25.

[31]   Eberl G, MacDonald HR. Rapid death and regeneration of NKT cells in anti-CD3epsilon- or IL-12-treated mice: a major role for bone marrow in NKT cell homeostasis. *Immunity* 1998;9:345-53.

[32]   Laloux V, Beaudoin L, Ronet C, Lehuen A. Phenotypic and functional differences between NKT cells colonizing splanchnic and peripheral lymph nodes. *J Immunol* 2002;168:3251-8.

[33]   Matsuda JL, Naidenko OV, Gapin L, Nakayama T, Taniguchi M, Wang CR, Koezuka Y, Kronenberg M. Tracking the response of natural killer T cells to a glycolipid antigen using CD1d tetramers. *J Exp Med* 2000;192:741-54.

[34]   Osman Y, Kawamura T, Naito T, Takeda K, Van Kaer L, Okumura K, Abo T. Activation of hepatic NKT cells and subsequent liver injury following administration of alpha-galactosylceramide. *Eur J Immunol* 2000;30:1919-28.

[35]   Taniguchi M, Nakayama T. Recognition and function of Valpha14 NKT cells. *Semin Immunol* 2000;12:543-50.

[36]   Gansert JL, Kiessler V, Engele M, Wittke F, Rollinghoff M, Krensky AM, Porcelli SA, Modlin RL, Stenger S. Human NKT cells express granulysin and exhibit antimycobacterial activity. *J Immunol* 2003;170:3154-61.

[37]   Carnaud C, Lee D, Donnars O, Park SH, Beavis A, Koezuka Y, Bendelac A. Cutting edge: Cross-talk between cells of the innate immune system: NKT cells rapidly activate NK cells. *J Immunol* 1999;163:4647-50.

[38]   Kitamura H, Iwakabe K, Yahata T, Nishimura S, Ohta A, Ohmi Y, Sato M, Takeda K, Okumura K, Van Kaer L, Kawano T, Taniguchi M, Nishimura T. The natural killer T (NKT) cell ligand alpha-galactosylceramide demonstrates its immunopotentiating effect by inducing interleukin (IL)-12 production by dendritic cells and IL-12 receptor expression on NKT cells. *J Exp Med* 1999;189:1121-8.

[39]   Hermans IF, Silk JD, Gileadi U, Salio M, Mathew B, Ritter G, Schmidt R, Harris AL, Old L, Cerundolo V. NKT cells enhance CD4+ and CD8+ T cell responses to soluble antigen in vivo through direct interaction with dendritic cells. *J Immunol* 2003;171 5140-7.

[40]   Cui J, Watanabe N, Kawano T, Yamashita M, Kamata T, Shimizu C, Kimura M, Shimizu E, Koike J, Koseki H, Tanaka Y, Taniguchi M, Nakayama T. Inhibition of T helper cell type 2 cell differentiation and immunoglobulin E response by ligand-activated Valpha14 natural killer T cells. *J Exp Med* 1999;190:783-92.

[41]   Singh N, Hong S, Scherer DC, Serizawa I, Burdin N, Kronenberg M, Koezuka Y, Van Kaer L. Cutting edge: activation of NK T cells by CD1d and alpha-galactosylceramide directs conventional T cells to the acquisition of a Th2 phenotype. *J Immunol* 1999;163:2373-7.

[42]   Zigmond E, Preston S, Pappo O, Lalazar G, Margalit M, Shalev Z, Zolotarov L, Friedman D, Alper R, Ilan Y. Beta-glucosylceramide: a novel method for enhancement of natural killer T lymphoycte plasticity in murine models of immune-mediated disorders. *Gut* 2007;56:82-9.

[43]   Eberl G, Brawand P, MacDonald HR. Selective bystander proliferation of memory CD4+ and CD8+ T cells upon NK T or T cell activation. *J Immunol* 2000;165:4305-11.

[44]   Nishimura T, Kitamura H, Iwakabe K, Yahata T, Ohta A, Sato M, Takeda K, Okumura K, Van Kaer L, Kawano T, Taniguchi M, Nakui M, Sekimoto M, Koda T. The interface between innate and acquired immunity: glycolipid antigen presentation by CD1d-expressing dendritic cells to NKT cells induces the differentiation of antigen-specific cytotoxic T lymphocytes. *Int Immunol* 2000;12:987-94.

[45]   Van Kaer L. Regulation of immune responses by CD1d-restricted natural killer T cells. *Immunol Res* 2004;30:139-53.

[46]   Miyake S, Yamamura T. NKT cells and autoimmune diseases: unraveling the complexity. *Curr Top Microbiol Immunol* 2007;314:251-67.

[47]   Novak J, Griseri T, Beaudoin L, Lehuen A. Regulation of type 1 diabetes by NKT cells. *Int Rev Immunol* 2007;26:49-72.

[48]   Miyake S, Yamamura T. Therapeutic potential of glycolipid ligands for natural killer (NK) T cells in the suppression of autoimmune diseases. Curr Drug Targets *Immune Endocr Metabol Disord* 2005;5:315-22.

[49]   Zajonc DM, Cantu C, 3rd, Mattner J, Zhou D, Savage PB, Bendelac A, Wilson IA, Teyton L. Structure and function of a potent agonist for the semi-invariant natural killer T cell receptor. *Nat Immunol* 2005;6:810-8.

[50]   Ulrichs T, Porcelli SA. CD1 proteins: targets of T cell recognition in innate and adaptive immunity. *Rev Immunogenet* 2000;2:416-32.

[51]   Gumperz JE, Brenner MB. CD1-specific T cells in microbial immunity. *Curr Opin Immunol* 2001;13:471-8.

[52]   Roark JH, Park SH, Jayawardena J, Kavita U, Shannon M, Bendelac A. CD1.1 expression by mouse antigen-presenting cells and marginal zone B cells. *J Immunol* 1998;160:3121-7.

[53]   Brossay L, Jullien D, Cardell S, Sydora BC, Burdin N, Modlin RL, Kronenberg M. Mouse CD1 is mainly expressed on hemopoietic-derived cells. *J Immunol* 1997;159:1216-24.

[54]   Brutkiewicz RR. CD1d ligands: the good, the bad, and the ugly. *J Immunol* 2006;177:769-75.

[55]   Zhou D, Mattner J, Cantu C, 3rd, Schrantz N, Yin N, Gao Y, Sagiv Y, Hudspeth K, Wu YP, Yamashita T, Teneberg S, Wang D, Proia RL, Levery SB, Savage PB, Teyton L, Bendelac A. Lysosomal glycosphingolipid recognition by NKT cells. *Science* 2004;306:1786-9.

[56]   Kinjo Y, Kronenberg M. Valpha14i NKT cells are innate lymphocytes that participate in the immune response to diverse microbes. *J Clin Immunol* 2005;25:522-33.

[57]   Mattner J, Debord KL, Ismail N, Goff RD, Cantu C, 3rd, Zhou D, Saint-Mezard P, Wang V, Gao Y, Yin N, Hoebe K, Schneewind O, Walker D, Beutler B, Teyton L, Savage PB, Bendelac A. Exogenous and endogenous glycolipid antigens activate NKT cells during microbial infections. *Nature* 2005;434:525-9.

[58]   De Silva AD, Park JJ, Matsuki N, Stanic AK, Brutkiewicz RR, Medof ME, Joyce S. Lipid protein interactions: the assembly of CD1d1 with cellular phospholipids occurs in the endoplasmic reticulum. *J Immunol* 2002;168:723-33.

[59]   Giabbai B, Sidobre S, Crispin MD, Sanchez-Ruiz Y, Bachi A, Kronenberg M, Wilson IA, Degano M. Crystal structure of mouse CD1d bound to the self ligand phosphatidylcholine: a molecular basis for NKT cell activation. *J Immunol* 2005;175:977-84.

[60]   Sriram V, Cho S, Li P, O'Donnell PW, Dunn C, Hayakawa K, Blum JS, Brutkiewicz RR. Inhibition of glycolipid shedding rescues recognition of a CD1+ T cell lymphoma by natural killer T (NKT) cells. *Proc Natl Acad Sci U S A* 2002;99:8197-202.

[61]   Zhou D. The immunological function of iGb3. Curr Protein Pept Sci 2006;7:325-33.

[62]   Hansen DS, Schofield L. Regulation of immunity and pathogenesis in infectious diseases by CD1d-restricted NKT cells. *Int J Parasitol* 2004;34:15-25.

[63]   Skold M, Behar SM. Role of CD1d-restricted NKT cells in microbial immunity. *Infect Immun* 2003;71:5447-55.

[64]   Tupin E, Kinjo Y, Kronenberg M. The unique role of natural killer T cells in the response to microorganisms. *Nat Rev Microbiol* 2007;5:405-17.

[65]   Wilson MT, Johansson C, Olivares-Villagomez D, Singh AK, Stanic AK, Wang CR, Joyce S, Wick MJ, Van Kaer L. The response of natural killer T cells to glycolipid antigens is characterized by surface receptor down-modulation and expansion. *Proc Natl Acad Sci U S A* 2003;100:10913-8.

[66] Stanic AK, Park JJ, Joyce S. Innate self recognition by an invariant, rearranged T-cell receptor and its immune consequences. *Immunology* 2003;109:171-84.

[67] Van Kaer L. alpha-Galactosylceramide therapy for autoimmune diseases: prospects and obstacles. *Nat Rev Immunol* 2005;5:31-42.

[68] Taniguchi M, Harada M, Kojo S, Nakayama T, Wakao H. The regulatory role of Valpha14 NKT cells in innate and acquired immune response. *Annu Rev Immunol* 2003;21:483-513.

[69] Wilson MT, Singh AK, Van Kaer L. Immunotherapy with ligands of natural killer T cells. *Trends Mol Med* 2002;8:225-31.

[70] Bezbradica JS, Stanic AK, Matsuki N, Bour-Jordan H, Bluestone JA, Thomas JW, Unutmaz D, Van Kaer L, Joyce S. Distinct roles of dendritic cells and B cells in Va14Ja18 natural T cell activation in vivo. *J Immunol* 2005;174:4696-705.

[71] Yamamura T, Miyamoto K, Illes Z, Pal E, Araki M, Miyake S. NKT cell-stimulating synthetic glycolipids as potential therapeutics for autoimmune disease. *Curr Top Med Chem* 2004;4:561-7.

[72] Schmieg J, Yang G, Franck RW, Van Rooijen N, Tsuji M. Glycolipid presentation to natural killer T cells differs in an organ-dependent fashion. *Proc Natl Acad Sci U S A* 2005;102:1127-32.

[73] Brossay L, Chioda M, Burdin N, Koezuka Y, Casorati G, Dellabona P, Kronenberg M. CD1d-mediated recognition of an alpha-galactosylceramide by natural killer T cells is highly conserved through mammalian evolution. *J Exp Med* 1998;188:1521-8.

[74] Miyamoto K, Miyake S, Yamamura T. A synthetic glycolipid prevents autoimmune encephalomyelitis by inducing TH2 bias of natural killer T cells. *Nature* 2001;413:531-4.

[75] Yu KO, Im JS, Molano A, Dutronc Y, Illarionov PA, Forestier C, Fujiwara N, Arias I, Miyake S, Yamamura T, Chang YT, Besra GS, Porcelli SA. Modulation of CD1d-restricted NKT cell responses by using N-acyl variants of alpha-galactosylceramides. *Proc Natl Acad Sci U S A* 2005;102:3383-8.

[76] Gonzalez-Aseguinolaza G, de Oliveira C, Tomaska M, Hong S, Bruna-Romero O, Nakayama T, Taniguchi M, Bendelac A, Van Kaer L, Koezuka Y, Tsuji M. alpha -galactosylceramide-activated Valpha 14 natural killer T cells mediate protection against murine malaria. *Proc Natl Acad Sci U S A* 2000;97:8461-6.

[77] Gonzalez-Aseguinolaza G, Van Kaer L, Bergmann CC, Wilson JM, Schmieg J, Kronenberg M, Nakayama T, Taniguchi M, Koezuka Y, Tsuji M. Natural killer T cell ligand alpha-galactosylceramide enhances protective immunity induced by malaria vaccines. *J Exp Med* 2002;195:617-24.

[78] Ortaldo JR, Young HA, Winkler-Pickett RT, Bere EW, Jr., Murphy WJ, Wiltrout RH. Dissociation of NKT stimulation, cytokine induction, and NK activation in vivo by the use of distinct TCR-binding ceramides. *J Immunol* 2004;172:943-53.

[79] Parekh VV, Singh AK, Wilson MT, Olivares-Villagomez D, Bezbradica JS, Inazawa H, Ehara H, Sakai T, Serizawa I, Wu L, Wang CR, Joyce S, Van Kaer L. Quantitative and qualitative differences in the in vivo response of NKT cells to distinct alpha- and beta-anomeric glycolipids. *J Immunol* 2004;173:3693-706.

[80] Wilson MT, Van Kaer L. Natural killer T cells as targets for therapeutic intervention in autoimmune diseases. *Curr Pharm Des* 2003;9:201-20.

[81]   Kakimi K, Guidotti LG, Koezuka Y, Chisari FV. Natural killer T cell activation
         inhibits hepatitis B virus replication in vivo. *J Exp Med* 2000;192:921-30.

[82]   Chackerian A, Alt J, Perera V, Behar SM. Activation of NKT cells protects mice from
         tuberculosis. *Infect Immun* 2002;70:6302-9.

[83]   Stenger S, Hanson DA, Teitelbaum R, Dewan P, Niazi KR, Froelich CJ, Ganz T,
         Thoma-Uszynski S, Melian A, Bogdan C, Porcelli SA, Bloom BR, Krensky AM,
         Modlin RL. An antimicrobial activity of cytolytic T cells mediated by granulysin.
         Science   1998;282
         121-5.

[84]   Apostolou I, Takahama Y, Belmant C, Kawano T, Huerre M, Marchal G, Cui J,
         Taniguchi M, Nakauchi H, Fournie JJ, Kourilsky P, Gachelin G. Murine natural killer
         T(NKT) cells [correction of natural killer cells] contribute to the granulomatous
         reaction caused by mycobacterial cell walls. *Proc Natl Acad Sci U S* A 1999;96:5141-
         6.

[85]   Kawakami K, Tohyama M, Qifeng X, Saito A. Expression of cytokines and inducible
         nitric oxide synthase mRNA in the lungs of mice infected with Cryptococcus
         neoformans: effects of interleukin-12. *Infect Immun* 1997;65:1307-12.

[86]   Kawakami K, Kinjo Y, Yara S, Uezu K, Koguchi Y, Tohyama M, Azuma M, Takeda
         K, Akira S, Saito A. Enhanced gamma interferon production through activation of
         Valpha14(+) natural killer T cells by alpha-galactosylceramide in interleukin-18-
         deficient mice with systemic cryptococcosis. *Infect Immun* 2001;69:6643-50.

[87]   Kawakami K, Kinjo Y, Yara S, Koguchi Y, Uezu K, Nakayama T, Taniguchi M,
         Saito A. Activation of Valpha14(+) natural killer T cells by alpha-galactosylceramide
         results in development of Th1 response and local host resistance in mice infected with
         Cryptococcus neoformans. *Infect Immun* 2001;69:213-20.

[88]   Jahng AW, Maricic I, Pedersen B, Burdin N, Naidenko O, Kronenberg M, Koezuka
         Y, Kumar V. Activation of natural killer T cells potentiates or prevents experimental
         autoimmune encephalomyelitis. *J Exp Med* 2001;194:1789-99.

[89]   Terabe M, Berzofsky JA. Immunoregulatory T cells in tumor immunity. *Curr Opin
         Immunol* 2004;16:157-62.

[90]   Hong S, Wilson MT, Serizawa I, Wu L, Singh N, Naidenko OV, Miura T, Haba T,
         Scherer DC, Wei J, Kronenberg M, Koezuka Y, Van Kaer L. The natural killer T-cell
         ligand alpha-galactosylceramide prevents autoimmune diabetes in non-obese diabetic
         mice. *Nat Med* 2001;7:1052-6.

[91]   Hammond KJ, Poulton LD, Palmisano LJ, Silveira PA, Godfrey DI, Baxter AG.
         alpha/beta-T cell receptor (TCR)+CD4-CD8- (NKT) thymocytes prevent insulin-
         dependent diabetes mellitus in nonobese diabetic (NOD)/Lt mice by the influence of
         interleukin (IL)-4 and/or IL-10. *J Exp Med* 1998;187:1047-56.

[92]   Hauben E, Roncarolo MG, Nevo U, Schwartz M. Beneficial autoimmunity in Type 1
         diabetes mellitus. *Trends Immunol* 2005;26:248-53.

[93]   Trop S, Samsonov D, Gotsman I, Alper R, Diment J, Ilan Y. Liver-associated
         lymphocytes expressing NK1.1 are essential for oral immune tolerance induction in a
         murine model. *Hepatology* 1999;29:746-55.

[94]   Menachem Y, Trop S, Kolker O, Shibolet O, Alper R, Nagler A, Ilan Y. Adoptive
         transfer of NK 1.1+ lymphocytes in immune-mediated colitis: a pro-inflammatory or a
         tolerizing subgroup of cells? *Microbes Infect* 2005;7:825-35.

[95]   Shibolet O, Alper R, Zlotogarov L, Thalenfeld B, Engelhardt D, Rabbani E, Ilan Y. Suppression of hepatocellular carcinoma growth via oral immune regulation towards tumor-associated antigens is associated with increased NKT and CD8+ lymphocytes. *Oncology* 2004;66:323-30.

[96]   Jinquan T, Li W, Yuling H, Lang C. All roads lead to Rome: pathways of NKT cells promoting asthma. Arch *Immunol Ther Exp* (Warsz) 2006;54:335-40.

[97]   Lalazar G, Preston S, Zigmond E, Ben Yaacov A, Ilan Y. Glycolipids as immune modulatory tools. *Mini Rev Med Chem* 2006;6:1249-53.

[98]   Goni FM, Alonso A. Biophysics of sphingolipids I. Membrane properties of sphingosine, ceramides and other simple sphingolipids. *Biochim Biophys Acta* 2006;1758:1902-21.

[99]   Sonnino S, Mauri L, Chigorno V, Prinetti A. Gangliosides as components of lipid membrane domains. *Glycobiology* 2007;17:1R-13R.

[100]   100. Elstein D, Abrahamov A, Hadas-Halpern I, Zimran A. Gaucher's disease. *Lancet* 2001;358:324-7.

[101]   101. Adar T, Ben-Ami R, Elstein D, Zimran A, Berliner S, Yedgar S, Barshtein G. Aggregation of red blood cells in patients with Gaucher disease. *Br J Haematol* 2006;134:432-7.

[102]   102. Stanic AK, De Silva AD, Park JJ, Sriram V, Ichikawa S, Hirabyashi Y, Hayakawa K, Van Kaer L, Brutkiewicz RR, Joyce S. Defective presentation of the CD1d1-restricted natural Va14Ja18 NKT lymphocyte antigen caused by beta-D-glucosylceramide synthase deficiency. *Proc Natl Acad Sci U S A* 2003;100:1849-54.

[103]   103. Lalazar G PS, Lador A, Pappo O, Zolotarov L, Ilan Y. . alleviation of ConA immune mediated hepatitis via glycolipids: The role of beta versus alpha configuration in determining NKT lymphocyte distribution and the TH1/ TH2 paradigm. *J Hepatology* 2006;44:S237.

[104]   104. Sehgal PB. Plasma membrane rafts and chaperones in cytokine/STAT signaling. *Acta Biochim Pol* 2003;50:583-94.

[105]   105. Park YK, Lee JW, Ko YG, Hong S, Park SH. Lipid rafts are required for efficient signal transduction by CD1d. *Biochem Biophys Res Commun* 2005;327:1143-54.

[106]   106. Margalit M, Ghazala SA, Alper R, Elinav E, Klein A, Doviner V, Sherman Y, Thalenfeld B, Engelhardt D, Rabbani E, Ilan Y. Glucocerebroside treatment ameliorates ConA hepatitis by inhibition of NKT lymphocytes. *Am J Physiol Gastrointest Liver Physiol* 2005;289:G917-25.

[107]   107. Ilan Y, Ohana M, Pappo O, Margalit M, Lalazar G, Engelhardt D, Rabbani E, Nagler A. Alleviation of acute and chronic graft-versus-host disease in a murine model is associated with glucocerebroside-enhanced natural killer T lymphocyte plasticity. *Transplantation* 2007;83:458-67.

Index

D

E

I

M

N

S